BISON
BOOKS

SEASONS

OF THE

TALLGRASS

PRAIRIE

A Nebraska Year

PAUL A. JOHNSGARD

UNIVERSITY OF NEBRASKA PRESS
LINCOLN AND LONDON

Library of Congress Cataloging-in-
Publication Data

Johnsgard, Paul A.
Seasons of the tallgrass prairie: a
Nebraska year / Paul A. Johnsgard.
pages   cm
Summary: "A collection of essays on
prairie wildlife and ecology"—Provided
by publisher.
Includes bibliographical references.
ISBN 978-0-8032-5337-7 (paperback: alk. paper)
ISBN 978-0-8032-5697-2 (epub)
ISBN 978-0-8032-5698-9 (mobi)
ISBN 978-0-8032-5696-5 (pdf)
1. Prairie ecology—Nebraska. 2. Natu-
ral history—Nebraska. 3. Nebraska—
Environmental conditions. I. Title.
QH105.N2J635 2014
577.4'409782—dc23
2014017095

Set in Adobe Garamond Pro and
Scala Sans Pro by Lindsey Auten.
Designed by Karla Johnson.

Dedicated to all who fought all the past battles to preserve and protect Nebraska's natural resources, to those doing so today, and to any who will take on the future challenges to keep Nebraska a special place for both humans and all of our fellow travelers.

CONTENTS

PREFACE AND ACKNOWLEDGMENTS

This collection of essays had its origins mostly by chance. At times, after a wonderful day out-of-doors in Nebraska, I have felt compelled to sit down and summarize some of my immediate past experiences. That is how the first of the essays came about, after a quasi-religious visit to an ancient Pawnee sacred site. Others, such as the essays on snow geese and the Platte River, were written only after months of growing concern over what I have come to believe is an increasingly short-range attitude about the value, beauties, and needs for preservation of our state's finite natural resources of land, water, and ecosystems. Still other essays were written at the suggestion of friends or to comply with a magazine or newspaper editor's request for a timely story.

In any case, nearly this entire collection of essays has, in large part, been extracted from my already published writings. The great majority of them were written for *Prairie Fire*, a monthly independent newspaper published in Lincoln. The progressive stance of *Prairie Fire* as to important environmental and political issues is so refreshing and welcome that I have happily complied with any suggestions by its editor, Cris Trautner, for submissions and have at times pestered her to accept still others. I also greatly appreciate her help in providing me with edited copy of all the essays that were first published in *Prairie Fire*, and the willingness of the newspaper's publisher, W. Don Nelson, to let me reproduce them. The original essays and all my other pieces I have published there can be found on the newspaper's website: www.prairiefirenewspaper.com.

Other than the *Prairie Fire* articles (which can be easily identified by their same or similar titles in the bibliographic sources section), I have, with permission, extracted parts or used all of three stories previously published in *Nebraska Life* magazine. These include the account of reproduction in the yucca and yucca moth (from "The Ancient Romance of the Yucca and the Yucca Moth"), the section of the prairie grouse essay that describes the interactions of a sharp-tailed grouse and prairie-chickens on a joint display ground (from "A Dozen Squaretails and a Sharpy"), and the descriptions of native grasses in an essay on tallgrass prairies (from "Autumn on the Prairie: Nebraska's Grasses"). Additionally, I have extracted some historical information on irrigation and corn production in the Platte Valley from my book *The Platte: Channels in Time* (University of Nebraska Press, 2008). In all cases, there has been some trimming, updating, or other modifications as has seemed desirable. The final essay is entirely new, and I appreciate the advice that Jim Douglas and Scott Taylor of the Nebraska Game and Parks Commission provided in fact-checking my comments about that agency.

Dr. Karine Gil-Weir kindly agreed to collaborate with me in writing two of the essays; her work at the Crane Trust has provided an important baseline for long-term population studies of both sandhill and whooping cranes in Nebraska and elsewhere. I also owe the Crane Trust thanks for letting me use their bunkhouse on many occasions, as well as using their crane blinds, and the same is true for the Rowe Audubon Sanctuary and the Nature Conservancy. And I would be remiss not to mention Tom Mangelsen and the entire Mangelsen family, whose cabin on the Platte River has often seemed like a second home to me and whose hunting blinds converted easily to photographic blinds, allowing Tom and me to often ruminate about the fate of the cranes, the Platte, and the natural world, and, while thus engaged, to often miss out on great photographic opportunities.

It is impossible to acknowledge all of the help I have directly or indirectly had in being able to write these pieces—they have grown out of a half century of roaming Nebraska's back roads, trails, and half-forgotten places among our grasslands, forests, rivers, and wetlands. Writing these essays has brought back a host of memories of locations, events, and golden days afield with hundreds of students, friends, colleagues, and others. They will know who they are.

Seasons of the Tallgrass Prairie

PART ONE

Wild Places and Natural Treasures

Sanderlings

1

A Place Called Pahaku

There is an area in eastern Nebraska where the Platte River, after flowing northeastward from the vicinity of Kearney for nearly 150 miles, enters the glacial drift bordering the Missouri Valley and turns directly east. Over its eastward course of about fifty miles, the river forms a shallow and wide sandy channel that is bounded to the south by forested bluffs and to the north by a wide, wooded floodplain. One of these glacially shaped and loess-capped bluffs was known historically to the resident Pawnee tribe as Pahaku (usually but incorrectly spelled as Pahuk) Hill. This Pawnee word may be roughly translated as "mound on or over water," or "headland." The bluff is one of five natural sites (four of them along the Platte River) in the historic range of the Pawnees that were considered sacred to them and is the only remaining location that is still virtually biologically intact. About fifty thousand years ago, during late glacial times, this bluff also marked the approximate point where the Platte River abruptly turned southeast. It then followed a glacial moraine valley, now known as the Todd Valley, toward present-day Ashland. Although this part of the lower Platte Valley is now recognized for its uncommonly rich bottomland soils, it is also rich in Pawnee history, since the Platte and Loup Valleys were among the most important parts of the Pawnees' original homeland.

Pahaku Hill is located almost directly north of Cedar Bluffs, in northern Saunders County. According to one Pawnee legend, a young boy once lay at the edge of the bluff, hoping to shoot a bird with his bow and arrows. Growing at the edge of the bluff was a tall cedar tree, marking the entrance to a huge cave that was the lodge of many animals. Several eagles and a hawk sat on the cedar tree, perhaps serving as guardians. A second, underwater entrance to the cave was also believed to exist, which could be reached only by following a kingfisher as a guide. The chief of the animals living in the lodge was a giant beaver, but the lodge also was the home of other spiritually important animals, such as deer, elk, antelope, wolves, coyotes, foxes, cranes, and geese.

These were known to be sacred animals (Nahu'ac) by the Pawnees, and in this cave they periodically held council. There they also endowed the young Pawnee boy with special healing powers, which he later passed on to others of his village. At times such medicine men visited Pahaku to renew their healing abilities and to give thanks. Of all the Pawnee animals having spiritual powers, birds were especially important. They served as direct messengers to the gods and played significant roles in important Pawnee ceremonies. Eagles were the most preeminent and powerful of these totemic birds, and hawks were also notable, as were their feathers. Owls were particularly significant in Pawnee healing ceremonies, while other species such as jays, magpies, and woodpeckers were appreciated for their own valuable attributes. For example, the intelligent magpie helped the legendary Pawnee child find the entrance to the Pahaku cave. There probably once was an actual cave at this site, as several of the Pawnees' sacred sites along rivers consisted of bluffs with caves, but erosion no doubt destroyed it long ago.

In a different and perhaps more authentic version of the legend, after a young boy had been sacrificed by his father and placed in the Platte River, two turkey vultures delivered his body to the sacred animals in the cave. The sacred animals brought him back to life and taught him all of their medicine powers. He later went back to his people to serve as a great medicine man and transmit his knowledge.

The Pawnees' peaceful bison-hunting and agricultural culture was eventually destroyed by the impact of European immigration, partly

·

as a result of the destruction of their bison-dependent economy. Their vast homeland, which once stretched from the Niobrara to the Arkansas and Cimarron Rivers and numbered about twenty thousand people by 1820, was decimated by smallpox during devastating plagues during 1831 and 1832. Adding to this catastrophe, part of their land was sold at a pitifully small price to the United States in 1833. Later losses of ancestral Pawnee territory were associated with the Kansas-Nebraska Act of 1854 and the ceding of tribal lands for settlement by immigrants. The Pawnees were soon limited to a small reservation along the Loup River, now Nebraska's Nance County. Eventually even this tiny remnant of their homeland was lost to settlement pressures. In 1874 the last of the Pawnees (about two thousand) left Nebraska, when all the adults walked to a small reservation in Indian Territory, now Oklahoma. At this time they were under periodic danger from attack by the Lakotas and were being increasingly surrounded by white settlements. After their reservation school year was over, the Pawnee children were similarly moved to Indian Territory. According to Pawnee oral history, they too walked the entire distance; after their shoes and moccasins had worn out they had to walk barefoot, with some dying along the way. The Pawnee reservation now consists of about twenty thousand acres, and the population near the end of the twentieth century consisted of about twenty-four hundred Native Americans, or about one-tenth of the estimated presettlement number.

Pahaku was homesteaded in 1868, and it was not until the forested part of the bluff was purchased by Dr. Louis and Geraldine Gilbert in 1962 that any attention was given to preserving its natural habitats. After learning of the location's great spiritual significance, the Gilberts applied to have the site placed on the National Register of Historic Places, which was approved in 1973. During the 1980s their land was preserved for posterity through a conservation easement. Later the Gilberts' land was sold to Kirby and Mary Zickafoose, who are equally determined to keep it in a natural and protected state. The Pat and Nancy Shanahan family have farmed the remainder of the bluff for more than a century, and in 2008 a delegation of Pawnees from Oklahoma visited Pahaku to help celebrate the establishment of a conservation easement on the Shanahans' farm, protecting that part of the bluff from further development.

Because of Pahaku's history and transitional location, linking the eastern deciduous forest plants and the prairie riverine forests, Ty Harrison, a University of Nebraska botanist, did an ecological analysis of the site's plants in 1984. He found that several eastern deciduous forest trees (bitternut hickory, black walnut, and American linden) approach or reach the western edge of their Platte Valley distribution at Pahaku. There are also several eastern woodland vines (carrion flower, bristly greenbrier, eastern virgin's bower, and Virginia creeper) and many woodland wildflowers (jack-in-the-pulpit, columbine, pale touch-me-not, white snakeweed, and American bellflower) that have similar eastern forest affiliations and range limits. Farther to the west, the drier climate and absence of a shaded forest understory increasingly prevent these plants from thriving and reproducing. In these ways, Pahaku represents a kind of botanical eastern outpost that also supports a comparable array of eastern forest-adapted animals such as eastern fox squirrels and white-tailed deer.

On a cold morning in mid-April 2010 I drove to Pahaku with a friend to meet with its longtime caretaker and fierce protector, Cherrie Beam-Clarke. Cherrie also has served for three decades as interpreter of the land's natural and Pawnee history and is an educational speaker for the Nebraska Humanities Council. The American plums were then in full bloom along woodland edges, while the leaves of most of the forest trees were just unfolding. Newly arrived migrants from the tropics such as brown thrashers and eastern phoebes were establishing territories, while permanent residents such as red-bellied and downy woodpeckers were making their presence known with territorial drumming. We walked a trail to one of the higher points on Pahaku Bluff, an open area of prairie where the carcass of a white-tailed deer had provided food for wintering bald eagles. From the hill one can visualize the course of the old Pawnee Trail that paralleled the southern bank of the Platte River, leading to distant Morse Bluff on the western horizon.

Walking along the bluff's steep slope, we flushed a pair of wood ducks from the trees where they had no doubt been looking for a suitable nesting cavity. Wood ducks are another eastern species that, like red-bellied woodpeckers, has progressively moved west along the Platte Valley woodlands. In addition to the widespread and early blooming blue violet, we found a few examples of Dutchman's-breeches. This delicate eastern woodland flower is very near the western edge of its Nebraska range and at Pahaku is limited to the bluff's steep and shady north-facing slopes, where it often grows among mosses and ferns. Its pinkish flowers resemble baggy pants hanging upside-down from a clothesline, through the narrow "waist" of which bumblebees must pass so they can reach the pollen tucked away in its spurs. The other early spring wildflowers we most wanted to see, columbine and jack-in-the-pulpit, had not made their brief but beautiful curtain calls. Like many other deciduous forest plants they have evolved adaptations allowing them to bloom and be pollinated before most sunlight is cut off by the leafy summer canopy.

We did find another botanical goal: the oldest bur oak in the area, which no doubt was already an impressive tree when the Pawnees were still living peacefully along the Platte. The great oak is still producing a few acorns but is slowly dying; one of its largest lower branches had recently broken off and lay desolate on the ground. The tree's twisted shape reminded me of an ancient Pawnee holy man, lifting his arms in anguish toward the sky and lamenting the fate of his dispossessed people, who now live in a reservation over five hundred miles away from their homeland.

2

The Life and Hard Times of the Platte River

When I was a graduate student at Cornell University in the late 1950s two of my best friends were doing field research in the Platte Valley of Nebraska on hybridizing bird species pairs found there, such as orioles, grosbeaks, towhees, and flickers. At the end of each summer they would return to Ithaca with stories of the beauty of the Platte's riparian woodlands, its birds, and the natural glories of central Nebraska in late spring. I had grown up in eastern North Dakota but had never visited Nebraska, so I listened with quiet jealousy to their descriptions of the clear-flowing, sand-bottomed Platte, mentally comparing it with the sluggish and muddy Red River that was within easy walking distance of my home.

Later I spent two years on postdoctoral fellowships in England. During my second year there I happened to learn of a job opening at the University of Nebraska, and my memories of those vivid descriptions of the Platte Valley immediately returned. I knew little else of Nebraska but had heard that, except for North Dakota, it perhaps had the best waterfowl-breeding habitats of any state south of Canada. As a waterfowl specialist, my dreams came true when I was offered the job sight unseen.

Arriving on campus in the fall of 1961, I could scarcely wait to visit the central Platte Valley in spring. In March 1962 I drove out to Elm Creek, then turned south to cross the Platte. Suddenly the fields were alive with seemingly endless flocks of sandhill cranes. Like Dorothy landing in Oz, I found myself in a world transformed.

From that moment on I knew I would remain in Nebraska for the rest of my life and that cranes would become a constant leitmotif for me. I quickly learned that the Platte Valley's spring crane population had been essentially ignored and was entirely unrecognized as a major ornithological phenomenon. In addition, the seemingly uncountable numbers of Canada and white-fronted geese and the endless flocks of ducks that poured into the Platte Valley during March and April took me back to my days as a teenager, when I waded through hip-deep marshes to try photograph these beautiful birds as they migrated northward by the millions along the western edge of the Red River Valley.

In the early 1980s I finally decided to write a short book on the Platte River and its ecology. By then the river was increasingly in danger from various sources. Its broad channels were gradually disappearing and being replaced by shrubs and trees, partly as a result of an absence of the prairie fires that once kept the trees in check. More importantly, the smaller and shallower river channels were entirely disappearing, a combined result of dewatering effects of upstream surface diversions and local groundwater extraction for irrigation.

A Prairie River and Its Groundwater Reservoir

To understand the current state of the Platte River a review of some history is needed. Nebraska's natural resources districts (NRDs) were established in 1969 by the state legislature in a belated and half-hearted attempt to try to manage the state's groundwater resources. These multicounty districts were sensibly organized along watershed boundaries. However, it wasn't until 1975 that they were given authority to regulate groundwater use through the state's Groundwater Management Act, by controlling well spacing, establishing pumping limits, and proposing controlled-use areas, all being subject to oversight by the state of Nebraska's Department of Natural Resources (DNR).

The development of center-pivot technology during the 1950s had resulted in a profusion of center pivots by the 1970s and had made the irrigation of even fairly hilly ground feasible. In 1970 there were about seven thousand center pivots operating in Nebraska, but by 2005 there were about seventy-two thousand center pivots in the state. By 2005 there were also 7.2 million Nebraska acres irrigated by groundwater, and another million irrigated by surface water. Two-thirds of all the irrigated lands were owned by the state's seven thousand largest farms, some of them larger than ten thousand acres and many controlled by out-of-state interests.

The year 1999 marked the start of a decade-long statewide drought. Nebraska was then leading the nation in irrigated corn acreage, and by the middle of that decade corn had become the most heavily federally subsidized of all U.S. crops. Nebraska was then fourth in the nation in total federal crop subsidies received, with most of them going into the pockets of the largest farm operations.

In 2004, after the initiation of litigation by a surface-water user complaining of damages owing to excessive groundwater pumping, a new state law (LB 962) gave full recognition to the need for the state to conjunctively manage surface water and hydrologically connected groundwater. The law directed the DNR to annually review all the river basins of the state as to the status of their surface and groundwater conditions. By the time the 2004 law was fully in effect it was really too late to put, as it were, the water back. By 2006, 94.4 percent of our state's groundwater extractions (averaging some 7,420 million gallons per day) were being used for irrigation by seventeen thousand users who represented only about 1 percent of the state's population.

Between 1999 and 2006, during the first half of the drought period, groundwater levels declined an average of nearly six feet in the Central Platte NRD. Not only did state groundwater levels begin to decline seriously at this time over broad regions of Nebraska, but the state's largest reservoir, Lake McConaughy, also began to suffer. This huge reservoir was built during the 1930s to serve for irrigation needs and hydroelectric power. It has a maximum capacity of 1.9 million acre-feet and is the largest reservoir in the entire Platte River system.

Lake McConaughy is also the source of nearly all of central Nebraska's surface irrigation. The reservoir reached near-record high-water levels during the relatively wet years of the mid-1990s, but the 1999–2009 drought brought it down to equally historic lows within five years. Smaller than normal snowpacks in the mountains of Wyoming and Colorado have produced greatly reduced runoffs into the headwaters of the South Platte and North Platte Rivers for most years since about 2000. As a result, by 2004 Lake McConaughy was at a historic low level of 20 percent capacity, and 2005 was the fifth consecutive year in which the amount of water flowing into the lake was less than half its normal inflow.

Coincidentally, in 2007 the U.S. Department of Energy proposed replacing 30 percent of the nation's petroleum use with ethanol by 2030. It has been estimated that up to about a fourth of this total might be obtained by using corn to produce ethanol without seriously cutting into the nation's basic domestic food and livestock needs. Thus, about 60 billion gallons of ethanol per year would be needed to replace 30 percent of the nation's annual gasoline consumption of about 150 billion gallons, since ethanol generates only two-thirds the energy content of gasoline.

To achieve even a fraction of the government's 60-billion-gallon ethanol-production target by 2030 would require massive additional federal subsidies to agriculture and ethanol plant construction, since in 2007 only about 7.5 billion gallons of ethanol were produced. Spurred by the prospect of new federal support, corn reached record-high prices by 2007. As a result Nebraska farmers planted about nine million acres of corn in 2007, the most since 1936. The average dollar value of irrigated land in Nebraska reached $2,150 per acre by 2006, up 36 percent from 2000. By 2012 prime irrigated land in eastern Nebraska was selling for as much as $8,000 per acre.

After a summer of widespread drought, Nebraska's 2012 corn crop of 1.3 billion bushels was down only slightly from the previous year's record production. It averaged over 140 bushels per acre, harvested from 9.15 million planted acres. The average 2012 harvest weight was 58.8 pounds of corn per bushel, in spite of severe statewide drought conditions and an annual rainfall that over much of the state's corn-producing region was nearly 10 percent below average. Little to no water was present in the Platte from Columbus east to its confluence with the Missouri River during the summer and early fall of 2012, killing uncountable numbers of fish and other river-associated wildlife.

The Platte's Threatened and Endangered Species

In 1994 the U. S. Fish and Wildlife Service decided to impose certain restrictions on the Platte River's use in order to protect four nationally threatened or endangered species that use the basin's natural resources, requiring all large water users, primarily irrigators, to assure that sufficient water is available to protect the habitats of the whooping crane, least tern, piping plover, and pallid sturgeon. The habitats of three of these rare or endangered species are mostly concentrated in the central Platte Valley, while the pallid sturgeon uses only the downstream stretch close to the Missouri confluence.

As a result, a consortium of persons representing diverse interests in the Platte Basin's water and natural environments came together to try find a way to equitably share the Platte's waters among a host of competing interests. The result was a much-debated compromise negotiated over a nearly decade-long period. The Platte River Cooperative Agreement, or Platte River Recovery Plan, was initially approved in 1997 as a three-year planning guide. As a part of a related and long-negotiated relicensing agreement for Kingsley Dam in 1998, the reservoir's operators agreed to set aside 10 percent of the storable inflows of Lake McConaughy (averaging about one hundred thousand acre-feet in normal years) as an Environmental Account. This water would be released for maintaining wetland habitats of the central Platte Valley when needed.

Parties involved in developing the cooperative agreement for managing the overall Platte River Basin included representatives of the federal government, Colorado, Wyoming, Nebraska, in-state natural resources districts and irrigation districts, and various national and state environmental groups.

In essence, the Platte River Recovery Plan attempted to fulfill the basic requirements of the Endangered Species Act. Its primary goal was to add and restore twenty-nine thousand acres of additional wetland habitats in the central Platte Valley. Some ten thousand acres are scheduled for such acquisition in this region during the first thirteen years of the plan.

In addition to its basic habitat goal of acquiring ten thousand new acres of wetland habitat, the Platte River Recovery Implementation Program calls for the management, lease, or securing of an additional nineteen thousand acres of open channels or other riverine habitats in the same region. These include some lands already owned by the Crane Trust (until 2012 officially known as the Platte River Whooping Crane Habitat Maintenance Trust), the Nebraska Game and Parks Commission, the Nature Conservancy, and the National Audubon Society. Furthermore, annual shortfalls to U.S. Fish and Wildlife Service target flow rates in the Platte River are to be reduced by 130,000–140,000 acre-feet through enhanced upstream storage in several reservoirs, such as Lake McConaughy, with the water to be released as needed. Although single annual spring pulses of water are the historically desirable pattern for river flows, flow rates for hydroelectric power generation meant that the Central Nebraska Public Power and Irrigation District (CNPPID) and the Nebraska Public Power District (NPPD) had to agree to modify their flows in such a way as to minimize harm to the three federally listed bird species.

To cover associated costs, the federal government has agreed to pay for as much as half of the approximately $320 million program, with the three states contributing the remainder, through a combination of cash, land, and water. Nebraska's share of the total cost is to be provided by land and water contributed by the NPPD and the CNPPID. However, Nebraska is also to be required to reduce existing stream-flow depletions to the 1997 level in order to offset water "depletions" resulting from irrigation wells drilled after approval of the initial 1997 preliminary agreement and before the final multistate approval of the Platte River recovery plan a decade later.

Irrigation interests offered endless objections to the Platte River plan, with four of Nebraska's natural resources districts firmly opposed to it. The Central Platte NRD even considered initiating a lawsuit that would attempt to remove the whooping crane from the list of federally endangered species. However, it was made clear to all the irrigation interests that, should they fail to agree to the plan's terms, any number of federally funded projects, such as dams, reservoirs, and hydroelectric plants, would receive close scrutiny to make certain that current activities did not jeopardize any endangered species or their habitats.

With the threat of expensive environmental surveys looming and potentially crippling alterations possibly required of their activities, the irrigation interests finally reluctantly agreed to comply. In the fall of 2006 all three governors signed on, and the secretary of the interior also added his approval a few weeks later, initiating a thirteen-year Platte River Recovery Implementation Program that will terminate in 2019.

The Platte River Recovery Implementation Program

The Platte's eighty-mile "Big Bend" stretches between Lexington and Chapman. It is nationally recognized as a critical habitat for whooping cranes, which have increasingly roosted on restored stretches of the river since the start of restoration activities about twenty years ago and have shown an encouraging if painfully slow long-term upward national trend.

The Platte Valley is also an important breeding area for both the nationally endangered interior race of the least tern and the threatened piping plover. Both species have responded rapidly to improved water-level management and the establishment of new nesting areas, such as at Rowe Audubon Sanctuary near Gibbon. As of the mid-1990s, Nebraska contributed about 20 percent of the piping plover's northern Great Plains population of about 1,250 pairs. By 2009 about 140 pairs of piping plovers were nesting along the Platte River, and in 2006 this segment comprised over 7 percent of the interior least tern's total breeding colonies (Mary B. Brown, personal communication). The restored areas of the Platte are also the single most important segment for seasonal use by waterfowl, shorebirds, and other migratory birds. By 2006 expanded ecotourism in the Platte Valley was generating an estimated $50–$100 million in annual total economic benefits.

Although not endangered, sandhill cranes have benefited greatly from increased protection and habitat restoration along this stretch of the Platte Valley. Since the 1980s sandhill crane numbers have gradually leveled off at five hundred thousand to six hundred thousand birds using the central Platte Valley each spring, making it by far the largest congregation of cranes anywhere in the world, and an influx of some two million snow geese has only increased the spring migration spectacle for birds, attracting fifteen to twenty thousand bird-watchers annually from around the world.

At least six thousand wetland acres in Nebraska had been acquired for restoration by purchase or from the Recovery Implementation Program participants by 2010, including a 2,650-acre ranch located between Elm Creek and Overton. This site, CNPPID's Cottonwood Ranch, has two miles of channel frontage on both sides of the river and is adjacent to lands already owned by the Nature Conservancy and the Crane Trust. Seven other major land acquisitions situated between Overton and Gibbon, ranging from 200 acres to 718 acres and totaling nearly 3,500 acres, have been identified as Platte River Recreational Access lands. Limited access for fishing, hiking, bird-watching, and mushroom-collecting, as well as hunting for deer, turkey, small game, and even waterfowl, are allowed there.

Yet the joint decision by the Platte River Recovery Implementation Program and the Nebraska Game and Parks Commission to allow late-winter and other hunting of waterfowl along this stretch of the river from the start of the year until March 23 is an ironic and counterintuitive use of land acquired in large part to protect whooping cranes. Although a 2006 Fish and Wildlife Service biological position paper identified the species' typical migration periods through Nebraska as running from March 18 to April 20 in spring and in fall from October 17 to November 10, whooping cranes are known to have arrived in the Platte Valley as early as February 10, and since the late 1970s one or two have regularly arrived with sandhill cranes before the end of February. Additionally, nearly half a million sandhill cranes are now regularly present in the Platte Valley by mid-March. Even if no whooping cranes are mistakenly shot under this arrangement, disturbance to roosting sites of sandhill cranes and other protected species by hunting activities might be severe.

The entire Platte River Recovery Implementation Program is to be evaluated for renewal in 2019. Additionally, the governor of any of the three cooperating states may decide to unilaterally withdraw his or her state from the program at any time. This is not an impossible scenario, given the decade-long foot-dragging that occurred among all of the three participating states before they signed on to its provisions.

In spite of this multistate and federal agreement that was so arduously cobbled together, the future of the Platte River lies primarily in the hands of Nebraskans, who, if past history can provide an example, have been inclined to let agriculture set the ground rules for water use. It would be a permanent stain on our society's values to let the river literally be sucked dry. The resulting silence of the cranes and waterfowl, and the sullied spirit of the hardy pioneers who followed the Platte westward, drank from its waters, and used its fertile soils and abundant water to become a prosperous and modern state, would represent the actualization of a vision that should ever haunt us.

3
Nebraska's Magical Spring Migration

Over the past fifty years I have often been asked why, of all the places I have lived and visited, I chose Nebraska as the place where I have decided to spend the rest of my life. I quite willingly admit that Nebraska lacks the mountain grandeur of Colorado, the wonderful rocky coastline of Oregon, and the stunning glaciers of Alaska. Yet I quickly point out that we Nebraskans can claim the continent's largest remaining native prairie flora and its associated prairie wildlife, perched on the largest region of picturesque sand dunes in the western hemisphere. This in turn rests atop one of the greatest reservoirs of fresh water in the world, whose artesian springs give birth to such beautiful Sandhills streams as the Calamus, Loup, Dismal, and Elkhorn. Then, as a trump card, I say, "Oh yes, and for six weeks in spring we also have what is one of the largest and most spectacular concentrations of birds in the world."

Compared with other spectacles I have seen, such as the great coastal seabird colonies of Alaska's Pribilof Islands, Nebraska's cranes and migratory waterfowl can be observed without fighting foul weather and high winds. The equally famous spectacle of at least a half million wildebeests migrating slowly across the Serengeti plains of Tanzania can be relatively boring and is also somewhat smelly. By comparison, there is boundless joy in sitting quietly among prairie grasses along the Platte River with an azure sky overhead and a chorus of crane music drifting down from a thousand feet above. As the silvery-gray birds wheel gracefully about in a giant vortex of life and call excitedly to one another as they descend to their safe and ancestral resting places in the river, I too know I have witnessed my personal Elysium.

There is a line in the movie *Field of Dreams* in which the ghost of the famous baseball player "Shoeless" Joe Jackson appears and, finding himself surrounded by a tall cornfield, says, "Is this heaven?" The response by the astonished landowner (Kevin Costner) was simply, "No, this is Iowa." Every March, when I reach the vicinity of Grand Island while heading west from Lincoln on I-80, I too silently ask myself, "Is this heaven?" By then the sky becomes increasingly sprinkled with skeins of countless geese overhead, the fields of corn stubble are progressively covered with foraging sandhill cranes, and the crisp but gentle south wind is redolent with the promises of an oncoming spring. Between Grand Island and Kearney the Platte Valley is an ever-changing spectacle of cranes, geese, ducks, and other migrating birds, ranging from undulating flocks of red-winged blackbirds resembling giant apparitions as they drift gracefully above the land to lone raptors, such as rough-legged hawks and Harlan's hawks, slowly working their way back toward their arctic breeding grounds.

Migrating waterfowl move into the Platte Valley from as early as mid-February in recent years, when the Platte River becomes ice-free and snowmelt in the nearby Rainwater Basin replenishes the shallow basins with enough water to provide resting and foraging sites for a dozen or more species of ducks and geese.

Even before the sandhill cranes have arrived in large numbers, ducks such as the common mergansers and common goldeneyes are likely to be seen forging along the ice-free opening of the river, and both mallards and northern pintails are likely to be seen coursing overhead in dizzying courtship flights. Soon they are supplemented by other dabbling ducks, including green-winged teals, American wigeons, gadwalls, northern shovelers, and diving ducks such as redheads, ring-necked ducks, lesser scaups, and canvasbacks. Among the last to arrive are blue-winged teals, finally back from widely scattered tropical wintering areas, and the tardy ruddy ducks, whose migrating males are usually still in their drab winter plumages.

There are few places in North America that, at least for a few weeks in March, support as much avian biomass as the central Platte Valley. Assume that each of the five hundred thousand sandhill cranes in the region weighs an average of six pounds and that snow geese are of similar weight. Between 2001 and 2003 there were estimations of from 1.2 million to 7.3 million snow geese, and a few thousand of the smaller Ross's geese, in the Rainwater Basin and central Platte Valley, averaging 3.2 million for the three years. Add to this a million or more Canada, cackling, and greater white-fronted geese that average perhaps five pounds. There are also several million mallards and northern pintails averaging about three pounds, and unknown numbers of other ducks of varied weights, and the total avian biomass must easily exceed fifteen million pounds.

These numbers are far too large to comprehend, but the amount of food required to maintain the flocks is even more mind-boggling. Suffice it to say that, without the great quantities of unharvested grain left over in the nearby fields, such ornithological richness would be impossible. It was the great agriculture revolution following World War II, with new technology providing machinery, fertilizers, pesticides, and new irrigation techniques, that altered the historic Platte Valley of small farms and precipitation-dependent crops into an agricultural powerhouse. No wonder the cranes and geese found that the Platte Valley was the place to go to put on a maximum amount of fat in a minimum amount of time prior to migrating to high-latitude breeding grounds, where food is relatively scarce.

Nebraska provides North America's most important spring staging grounds for both sandhill and whooping cranes, and the central Platte Valley contains the most important of all the state's wetlands for both species. For the sandhill cranes, it is the combination of the Platte River as an ideal roosting habitat and the nearby cornfields as a convenient source of easily gathered food. For the whooping cranes, the Platte is less attractive as a roosting habitat than it was historically, since so much of it has been overgrown with shoreline and island vegetation. The foraging niche of whooping cranes is more water-based than is the sandhills', with much of the whooping crane's traditional foods obtained from shallow ponds, marshes, or estuaries. Nevertheless, these endangered cranes still annually stop in the Platte Valley. They sometimes feed in cornfields along with sandhill cranes, but they are more often found in rather remote wetlands and riparian lowlands both south and north of the immediate Platte Valley.

We have no idea as to how long this magnificent spectacle has gone on. Cranes have waded the wetlands of the world for at least fifty million years, a period more than ten times longer than the time since ancestral humans first walked upright. They have probably migrated across what is now Nebraska for several million years, or longer than the Platte River has been in existence. It seems likely that, at the end of the Pleistocene epoch more than ten thousand years ago, as outwashes from the retreating glaciers of the northern plains were spilling into the Missouri River and the glaciers of the western mountains were also melting, the Platte River or its antecedents had their origin, carving new water routes through the tundra-like landscapes of the central plains. We can also imagine that sandhill cranes and other arctic-adapted birds such as snow geese annually followed the warming landscape northward, breeding during the brief summers and retreating to ice-free wetlands each winter.

No doubt early Native Americans saw and rejoiced in these flights; they must have represented a vital source of fresh meat after a winter of probable near-starvation. But the crane migration remained largely unknown until very recent times; early immigrants following the Platte west typically didn't cross Nebraska Territory until late spring, so that they might cross the mountain passes of Wyoming during the snow-free period of summer.

The first estimates of the crane migration date from the early 1940s. A writer in the April 19, 1934, *Hastings Daily Tribune* hypothesized that the stretch of the Platte between Kearney and Odessa "is crossed twice a year by more sandhill cranes than any other strip of similar length in the same latitude anywhere from coast to coast." In 1945 Dr. William Breckenridge, an ornithologist from the University of Minnesota, provided one of the first numerical estimates of spring crane numbers in the central Platte Valley. His counts suggest that in this period, before the advent of extensive use of fertilizers, pesticides, and irrigation, the numbers of sandhill cranes seen near the confluence of the North Platte and South Platte Rivers were in the general range of thirty thousand to forty thousand birds. No contemporary estimates of birds farther downstream are available, but one might imagine a regional population of no more than one hundred thousand or so birds existed then.

By the early 1960s, with the legalization of sandhill crane hunting in Texas and New Mexico, the U.S. Fish and Wildlife Service began systematic surveys of spring sandhill crane numbers in the Plate Valley, where essentially all of the midcontinent arctic-breeding lesser sandhills and many of the subarctic-breeding "Canadian" sandhill cranes concentrate in March. Early survey estimates were subject to great variation but generally ranged from about 150,000–200,000 cranes. By the mid-1970s U.S. Fish and Wildlife Service data suggested a maximum Platte Valley population of 200,000–270,000 sandhills. Since then the population estimates for the Platte Valley have gradually crept upward, so that now 450,000–500,000 represents the generally accepted estimate. This estimate has remained fairly stable recently and has been generally supported by independent ground surveys and other methods of aerial survey.

It is difficult to know what factors might have prompted the remarkable growth of the midcontinent population of sandhill cranes during the past seventy years, but the massive increase in corn production in the Platte Valley since the 1940s and the attendant availability of unharvested corn left in the fields each fall has certainly been important. This corn bonanza has provided a nearly unlimited source of food for cranes as they move north in the spring, when they need maximum fat storage to prepare them for the rigors of their remaining migration and the stresses associated with an arctic nesting environment. Quite possibly the amelioration of the climate in arctic regions has also had an important effect. By lengthening the frost-free period to a point that the nearly one hundred days required for nest building, egg laying, incubation, and chick rearing can now often be comfortably fit between the last spring blizzard and the first fall snows.

Even with a successful breeding season a pair of sandhill cranes only infrequently manages to produce and raise two chicks from their two-egg clutch to fledging. Besides other losses, one of the eggs (the first laid) always hatches a day or two before the other, and the first-hatched chick may fight with and might even kill its younger sibling. Should both young fledge, however, the family soon forms a strong social bond, which is held together by mutual calling and postural displays. A juvenile crane is likely to remain closely attached to its parents for up to about three years, when sexual maturity leads to independence. Many of the sandhill cranes breeding in the mild climate of Florida may attempt to breed when only two years (in males) to three years (in females) of age. However, it is believed that arctic-breeding cranes undergoing the stresses of long migrations and extreme climatic conditions may require up to five years in order to achieve breeding status.

How long social attachments among parents, siblings, and other close relatives persist is still uncertain in wild cranes. It is generally true that all species of cranes form permanent pair bonds, but these are often broken by the death of one of the partners or, less often, by divorce. Divorce is rare among sandhill pairs that have a history of successful reproduction, and the ability of the male to establish and maintain a desirable nesting territory seems to be a major factor in determining whether a pair bond will be maintained between seasons.

Family bonds among whooping cranes may be stronger than those in sandhills. Long-term observations of color-banded birds has shown that the small flock sizes typical of whooping cranes result from the long-term social attachments of closely related individuals, so that as many as four generations of birds can be present in a single migrating assemblage. This factor, along with the great potential longevity of cranes, often exceeding thirty years in protected populations, allows for real cultural transmission of information from generation to generation. The older birds may actively or passively pass on important information about migration routes and migratory staging areas as well as suitable and secure breeding and wintering sites.

Through such intergenerational learning, it is likely that an adult arctic-breeding sandhill crane knows more about the intimate geography of North America than does a professional airline pilot. Furthermore, each arctic-breeding crane must undertake these hazardous migrations of three thousand to four thousand miles or more twice yearly. These migrations are performed in the face of legal hunting in every Canadian province and almost every American state located on the birds' migration route between Alaska or Canada and Texas. Nebraska is the only Central Flyway state that has never attempted to obtain permission from federal authorities for initiating a sandhill crane hunting season.

Sandhill crane killing for "sport" has greatly increased in popularity since its initiation in Texas and New Mexico in the early 1960s; now roughly forty thousand cranes in the midcontinent population are killed each year, which is about 8 percent of the total, or most of each year's total production of juveniles. This annual hunting mortality rate means that a crane living in the Central Flyway has about a 50 percent greater danger of being killed during any single year at the hands of a hunter than did a U.S. soldier serving over the entire ten-year period of the Iraq War.

For those sandhill cranes that survive the long hunting season and winter months in the playa wetlands and marshes of Texas, New Mexico, Arizona, and Mexico, their safe return to the Platte River and its surrounding corn-rich valley can only be a vast relief. We humans too can also briefly forget our worries of daily life by visiting the Platte Valley in early spring and getting a glimpse of wild America. One need not be a bird expert to enjoy this almost incredible visual and auditory experience, which I can promise will indelibly last for a lifetime.

Visitors to the central Platte Valley may choose to experience a sunrise or sunset roosting flight in a commercial blind, such as at Audubon's Rowe Sanctuary near Gibbon, or one operated by the Crane Trust's Nature and Visitor Center at the I-80 Alda interchange near Grand Island. Or one may simply watch the amazing passing parade of cranes, geese, and other birds from one of the two public viewing platforms, which are located at the Platte River bridges along the highways directly south of Alda and Gibbon. In any case, don't pass up a sunrise or sunset experience in favor of a quick trip; to do so is to cheat yourself out of a brief glimpse of nirvana.

4

The Birds of Nebraska's Boondocks

The first time I set foot on the shores of Lake McConaughy in the mid-1970s, it was against my will. I had been asked to teach ornithology at the University of Nebraska's newly established Cedar Point Biological Field Station and anticipated enduring a several-week stay at a hot and mosquito-rich location. I was still grumbling to myself as I approached Kingsley Dam, about nine miles north of Ogallala, and turned off on a narrow gravel road leading through a steep, rocky canyon. Suddenly I flushed a great horned owl from its nest, startled a magpie, and could hear rock wrens singing from the canyon walls. In an instant my mood shifted to elation, and I began one of the happiest experiences of my life.

For seventeen years I taught at Cedar Point, getting the background I needed for writing a book on the Sandhills region and, for the first time, encountering the avian wonders of western Nebraska. Besides the proximity of Lake McConaughy, and its overflow wetland Lake Ogallala, the area offered the wet meadows and marshes of the vast Sandhills region immediately to the north of the two lakes. Before Kingsley Dam was modified during the 1980s to produce hydroelectric power, Lake Ogallala was little more than a deep marsh with fairly stable water levels. It had acres of cattails providing a nesting habitat for sora and Virginia rails, marsh wrens, ruddy ducks, and coots and a shoreline of willows that historically had nesting American bitterns and a colony of black-crowned night herons. The evening chorus of rails, along with the soft calls of poorwills and great horned owls from the nearby canyons, provided perfect evening vespers.

After eight years at the station, and with the great help of Richard Roche, I assembled the area's first checklist in 1984. That preliminary list contained 244 species, establishing the Lake McConaughy region as a birding hotspot in the state, even exceeding Crescent Lake National Wildlife Refuge in its species diversity. Later additions to the list in 1996 raised the total to 305 species. These included 104 breeding species and 17 more that were probably breeding. Additional species, bringing the total to 342, were added by Charles R. and Mary Bomberger Brown in 2001. Most recently, the total has been further raised to 363 species in a publication by the Browns and Steven Dinsmore, making the region one of the most bird-rich localities in all of interior North America.

Lake McConaughy and Lake Ogallala collectively offer Nebraska's largest aquatic habitats, with about 105 miles of shoreline and wetlands. Lake McConaughy is about 22 miles long, 3 miles wide, and up to 142 feet deep and has at least thirty thousand acres of water surface at full capacity. The two lakes attract a great variety of Nebraska's waterfowl (thirty-seven species), shorebirds (thirty-six species), and gulls (seventeen species). They also attract the state's greatest concentration of bald eagles during late fall and winter, when birders can watch them from a free-access viewing building located just below Kingsley Dam. Lake McConaughy's Christmas Bird Count always produces the largest species total in the state and the highest total anywhere in the Great Plains north of Texas. During the decade 2000–2001 through 2009–2010, the average number of species seen was 96, with a record high count of 109 species in 2010–2011. By comparison, the Christmas Counts for Lincoln, Nebraska, over that same period averaged 62 species.

The Lake McConaughy Christmas Count always generates rarities. During the ten-year period just mentioned, there were single-year sightings of Pacific loon, northern goshawk, little gull, black-headed gull, Inca dove, mountain chickadee, and gray-crowned rosy finch. There were sightings during two years of the tufted duck and Bohemian waxwing. The Barrow's goldeneye, white-winged scoter, red-necked grebe, and mew gull were seen during three counts, and the great black-backed and Iceland gulls were observed in four. Collectively, a remarkable twelve species of gulls and thirty-one species of waterfowl were tallied during the Christmas Counts of that decade. At least sixteen species that are fairly regularly seen in Nebraska during warmer seasons, but are almost always gone from the state by late fall, were also found during these counts.

The peak of spring migration, and the best time for seeing the largest number of species, comes in mid-May, the peak arrival period for warblers and other insect eaters. The last species to arrive, by late May, are the yellow- and black-billed cuckoos, poorwill, and common nighthawk. From 1992 to 1997 Bill Scharf and his assistants banded over eleven thousand birds in the Cedar Point vicinity. Over three hundred individuals each of eleven species were banded. In order by descending number of birds banded, the eleven species were the orchard oriole, red-winged blackbird, house finch, American goldfinch, yellow warbler, cliff swallow, house wren, common yellowthroat, chipping sparrow, lark sparrow, and yellow-breasted chat. Additionally, the area is in the contact zone of several east-west pairs of closely related birds that interact and sometimes hybridize. Banding at Cedar Point Station has revealed hybrids between indigo and lazuli buntings, spotted and eastern towhees, and Bullock's and Baltimore orioles. The rose-breasted and black-headed grosbeaks are also present and are known to hybridize elsewhere along the Platte River system. These observations prove that this area is a true transition zone between eastern and western faunas.

From 1982 through 2011, over 210,000 cliff swallows were banded and studied by Charles R. and Mary Bomberger Brown and their assistants, perhaps the largest number of any songbird species ever banded in North America, and an indication of the remarkable abundance of cliff swallows in this region. Lake McConaughy is also one of the Great Plains' most important nesting grounds for the nationally threatened piping plover and the nationally endangered interior least tern. It is also one of the few known nesting sites in the state for the Clark's grebe, a close relative of the more common western grebe.

If this array of birdlife were not enough, it is only some seventy miles to Crescent Lake National Wildlife Refuge, a forty-one-thousand-acre refuge in the midst of the Sandhills. It is located along the transition zone between the semi-alkaline wetlands of the central Sandhills and the hyper-alkaline wetlands of the westernmost ones. The refuge's many shallow wetlands are a prime migratory stopover site for myriad shorebirds. Upland sandpipers, willets, and long-billed curlews also nest here. The Crescent Lake refuge is perhaps the best single location in Nebraska for seeing nesting waterfowl and, because of its alkaline wetlands, it attracts a few distinctly rare western species such as the cinnamon teal. It also hosts other equally attractive breeding birds, including the ruddy duck, redhead, canvasback, western grebe, and black-crowned night heron. Forster's and black terns also breed here, as well as colonies of eared grebes and double-crested cormorants. A current refuge checklist lists 291 species, including at least 85 nesting birds.

One of my favorite memories of the Crescent Lake refuge occurred on an early May morning in the 1980s when the refuge was awash with migrating sandpipers. I had stopped my car and was gazing at them when a peregrine falcon flashed across my view for an instant, and the shorebirds suddenly scattered like exploding confetti. It happened so fast I didn't have time to grab my camera and try to document the event, but it is imprinted on my memory as clearly as any photograph could have recorded it. I also have cherished memories of western grebes with chicks on their backs, of aerial bombardments by long-billed curlews while defending chicks hidden somewhere in the tall grasses, and of ruddy ducks courting, with the brilliance of their intensely blue bills rivaling that of the sky overhead.

In the more remote Sandhills of western Garden, southern Sheridan, and northeastern Morrill Counties the wetlands become increasingly alkaline, to the point that few shoreline or aquatic plants can survive, and the aquatic animal life is mostly limited to saline-tolerant invertebrates such as brine shrimp and brine flies. These are favorite foods of phalaropes, the whirling dervishes of the bird world, which spin in tight circles while swimming on the water surface, producing an upward vortex that brings these organisms to the surface, from which they can be easily plucked. Many of the Wilson's phalaropes that pass through in April and early May remain to nest, as do American avocets and occasional black-necked stilts. Vast flocks of Wilson's phalaropes and smaller numbers of red-necked phalaropes crowd into these marshes in early May, together with American avocets, black-necked stilts, white-faced ibises, and other shorebirds. The best of these alkaline marshes are on private lands, but driving south from Rushville to Lakeside provides a peek at some of them that are close to the road and a feeling for the beauty of the region. There may not be a real nirvana in this world, but at least for naturalists the Sandhills wetlands of Nebraska come very close.

5

What Is a Tallgrass Prairie?

I have come to love the color and majestic stature of big bluestem. I delight in letting my open fingers run through the grass
as I walk through head-high stands in autumn, until my shirt is
tinted and spattered golden yellow by its pollen. If there is a king
of American grasses, big bluestem must surely be it. Big bluestem
is a warm-season grass, doing most of its growth during the hot
summer months and finally bursting into blossom in September.
By October it is beginning to shed its seed crop. Its generic Latin
name, *Andropogon*, translates as "man's beard," a fine description
of its flowering head.

If big bluestem is the king of prairie grasses, Indiangrass, a
companion tallgrass species, would be an appropriate queen. The
bronzy fall color of Indiangrass is even more beautiful than big
bluestem. Its bushy head of autumn florets seems to me to resemble
a golden-tasseled wand. Indiangrass and big bluestem can grow to
six or more feet high in good years, and even to eight feet or more
in very moist years.

Little bluestem is Nebraska's "shaggy" prairie grass, of which Willa
Cather wrote lovingly in *My Ántonia*. Little bluestem's English name
refers to a bluish cast on the lower leaves and stem nodes. However,
by midsummer much of the entire visible plant is starting to turn
a rich reddish tint. By fall one can easily recognize little bluestem
by its bunch-like, shaggy shape and its wonderful overall coppery
red color, almost matching the colors of an autumnal prairie sunset.

Like big bluestem, it is a warm-season species, growing the most in the summer months and sending out graceful, feathery flowering stalks in early fall. Its abundant seeds are soon dropped, but the upright stems and leaves persist over the winter. In good years little bluestem may produce two hundred or more pounds of seeds per acre, providing important fall and winter food for small mammals and native birds.

Little bluestem is easily the most important plant of Nebraska's mixed-grass prairie. It not only vies with big bluestem for dominance in the eastern tallgrass prairie but also penetrates the entire Sandhills region. Buffalo grass was the food for immense herds of bison that once migrated through Nebraska each spring and fall. Buffalo grass is the classic short grass of western and northwestern Nebraska. It forms a mat-like turf at the ground, beginning its growth in late spring and continuing throughout the summer. To a greater degree than any of the other short grasses, it tolerates repeated grazing and is tough enough to withstand a variety of soil and climatic conditions.

Buffalo grass is unusual in that the sexes are on separate plants. Unlike the other prairie grasses, its seeds are a hard bur. Without treatment, such as chilling, soaking, or passing through the digestive tract of an herbivore, few of these seeds will germinate. This trait may have helped buffalo grass spread with the migrating bison from its geographic origins in Mexico north throughout western North America.

Although buffalo grass now occurs only in western Nebraska under natural conditions, it probably extended farther east during the drought years of the 1930s. Some cultivated varieties can grow well as far east as Lincoln, at least during drier summers.

At the center of Nebraska is the vast Sandhills region, where moist meadows and marshlands fed by water from the Ogallala Aquifer bring this region's plant diversity to more than four hundred species, but the most important upland Sandhills plants are perennial bunchgrasses. These include little bluestem, hairy grama, and Junegrass, which all grow to heights of two to three feet. The region also hosts tall grasses, such as sand dropseed, sandreed and sand bluestem, a sand-adapted relative of big bluestem, our king of prairie grasses.

To visit any prairie in Nebraska is, in a sense, to visit with our very distant relatives, each of which has its own story to tell, if we will only try to understand. Although each season is different, autumn is a very special time to visit a Nebraska prairie. Life has by then come full circle, and it is a perfect time to sit or lie down in the grass, to enjoy the sights, sounds, and smells of nature close at hand, and to at least briefly merge one's soul with our natural world.

When I was very young, I used to walk along railroad track right-of-ways near my home in the Red River Valley of eastern North Dakota. I didn't know that the "turkey-foot grass" that grew higher than my head was something special and that under its more formal name of big bluestem it is a charter member of the tallgrass prairie that once covered much of eastern North Dakota. Somewhat later my mother began to teach me about some of the native prairie flowers that grew in low meadows near their once-homesteaded farm near the Sheyenne River. Today this area has been preserved as part of the Sheyenne National Grassland, the largest federally owned area of tallgrass prairie in America. I learned there to identify such beautiful plants as tall blazing star and Canada goldenrod and acquired at least a nodding acquaintance with milkweeds, sunflowers, and some of the other more common and colorful wildflowers. At least as importantly, I learned to associate such glorious birds as marbled godwits and bobolinks with patches of native prairie, which even then were mostly confined to very hilly, very rocky, or very sandy sites at the very edges of or beyond what was once glacial Lake Agassiz, the heart of the Red River Valley. I much later learned that such relatively rare prairie plants and animals are "indicator species" and that if one wishes to find them (and protect them), it is necessary to protect the entire prairie community.

When I came to Lincoln in 1961 there were still dozens of relict prairies near town, where I could go to see the prairie birds and plants of my childhood. But as the years passed, these prairies were converted one by one to agriculture or suburban developments, like disappearing Cheshire Cats. But these disappearing cats usually didn't leave so much as a smile. A few scattered teeth were often all that remained, in ditches and at the edges of fields, where the deep roots of perennial grasses like big bluestem allowed them to continue for a time their losing battle against plows and herbicides. Even the fairly new house we bought at the then-outskirts of Lincoln still had a few shoots of big bluestem that fought valiantly for a few years against the socially acceptable bluegrass. After being warned by the authorities about tolerating such "weeds" in my yard, I too accepted defeat.

One of the few remaining public-access prairies near Lincoln persisting into the twenty-first century is Nine-Mile Prairie, a once privately owned pasture of some eight hundred acres on hilly glacial moraine that had been studied intensively during the 1930s by Professor John Weaver and his students. During World War II part of the prairie was taken over by the military for use as an ammunition storage site, and its size gradually diminished to somewhat more than two hundred acres. The prairie was acquired by the University of Nebraska in 1983 and is now protected and managed both for research and as a historic prairie. It is freely available to the public for nonconsumptive purposes, such as nature study, birding, and hiking.

By a stroke of good fortune, and some ambitious money-raising on the part of the state and local chapters of the National Audubon Society, a large tract of prairie was acquired in 1998 that is just three miles south of Denton and less than twenty miles southwest of Lincoln. This prairie is also located on unplowed glacial moraine and is almost as botanically diverse as Nine-Mile Prairie. Spring Creek Prairie Audubon Center is now the jewel in Nebraska's prairie crown, with a modern interpretive center and a staff including several trained biologists. Originally 626 acres in area, local fundraising has brought the prairie's total acreage to 840 acres, of which about 650 acres comprise native prairie.

The flora of Spring Creek has now been well inventoried, and it includes over 370 plant species. Excluding trees, aquatic plants, and woods-adapted plants, there are well over 200 prairie species. Although the majority of the individual plants are perennial grasses, grass species actually make up only about 20 percent of the native prairie floral species diversity. Broad-leaved herbs, which are collectively called forbs, constitute about 70 percent of the species, while shrubs and a few woody vines add the remaining 8 percent. So it is the forbs that give the greatest structural complexity to prairies, and these include a large number of plants in the sunflower (aster) family, fewer in the legume family, and very few in such families as the orchids, only one of which is known to occur at Spring Creek. While the grasses are wind pollinated, many of the forbs are pollinated by insects, and it is this latter adaptation that has produced the displays of multicolored and scented flowers in spring and summer that at times turn the tallgrass prairie into a garden. At Spring Creek Prairie the flowering of the purple coneflower in June and July is a summer highlight, while several asters, such as New England aster and other blue to purple asters, vie with downy gentians to be the final fall hosts to honeybees and bumblebees during late October.

I confess that spring is my favorite time to visit Spring Creek, when the migratory birds are returning and the first spring flowers, such as violets, rush into bloom to complete their flowering before being shaded out by the earlier grasses and taller forbs. But each season has its attractions. The tall prairie grasses are nearly all "warm-season" species, waiting for the oppressive heat of midsummer to put on their most rapid growth. By September the Indiangrass and big bluestem may easily exceed six feet in a wet year, and to lie down in a stand of these grasses and look toward the sky above is to get an ant's view of its world.

The tallest hills of Spring Creek Prairie are among the highest points in Lancaster County, affording a spectacular, unobstructed view in all directions. Sitting on one of these hilltops one can close one's eyes and listen to the sounds of near solitude, sometimes marked only by the songs of a distant western meadowlark, the scream of a soaring red-tailed hawk, or, in spring, the soft kettledrum sounds of courting greater prairie-chickens. In 2008 the spring equinox happened to fall almost exactly on the night of the full moon, so I decided to watch the simultaneous sunset and moonrise from the top one of these tall hills. I sat on a large quartzite boulder that protruded a few feet above the ground, a souvenir of the melting glacier that had shaped these hills during the last ice age. It was like watching one beautiful curtain fall in the west as another, equally stunning, curtain was rising in the east.

By November the prairie has quieted down, with the starches, sugars, and other carbohydrates that were manufactured by perennials during summer now safely stored in root systems many feet below ground, well out of the reach of grazing animals. What is left are the rusty brown skeletons of leaves and stems that make for spectacular fall panoramas, especially when contrasted with the red leaves of shrubby sumacs and the blues of cloudless fall skies.

Winter is a time for hardy souls to walk the prairie trails in search of snowy tracks marking the passage of coyotes, rabbits, deer, raccoons, mice, and other mammals that otherwise are likely to remain hidden. Much of the activity of small rodents occurs under deep snow; its insulating quality allows the temperature at ground level to remain only a few degrees below freezing even if the air temperature above the snow should approach zero. Red foxes, coyotes, and some owls can hear the sounds made by unseen mice and voles as they scurry about unseen and will suddenly pounce on them from above. By December the long blue shadows of grass cast on the snow by the pallid winter sun provide only cold comfort, but they do offer the promise of a sun that by January will be rising sooner, slowing increasing in strength and providing both life-giving light and heat to the waiting plants and animals.

6

Close Encounters with Nature at Spring Creek Audubon Prairie

Spring Creek Prairie shimmers like a newly woven copper-colored blanket in the brilliant sunlight of late October. Covering more than eight hundred acres of glacially sculpted land in southeastern Nebraska and located less than twenty miles southwest of Lincoln, its high hills represent the western limits of the last great glacier reaching this far south. Spring Creek's hilly ground is intermixed with rich soil materials carried in from the north and blown in from the west, but its undulating surface and rock-strewn substrate have protected it from the plowing and cropping that were the fate of nearly all of eastern Nebraska's fertile lands.

Many scattered boulders, transported by the ice from as far north as South Dakota and Minnesota, are visible near the tops of some of the highest hills, where ten thousand years of more recent erosion has gradually exposed them. Yet even the largest of these boulders become seasonally hidden by the tall perennial grasses after they have attained maturity in early fall. The tallest of these grasses, big bluestem and Indiangrass, grew as high as eight feet after the generous summer rains of 2011. These impressive grasses effectively hide most of the other 370 species of plants that have been identified in this relict prairie, one of the two largest remaining tallgrass prairies in all of Nebraska. Biologists have also documented at least 27 species of mammals, 220 birds, 53 butterflies, and 35 dragonflies and damselflies at Spring Creek.

Because of its diverse plant and animal life, Spring Creek Prairie Audubon Center has been a magnet for naturalists ever since the National Audubon Society acquired it in 1998. Its educational value was greatly enhanced with the completion of a beautiful interpretive center in 2006. Its hay-bale construction and low-slung architecture allow it to fit appropriately and inconspicuously into the prairie surroundings.

In recent years Spring Creek and Lincoln's Pioneers Park Nature Center have expanded their educational programs to encompass all of the fourth-grade children in Lincoln's public schools during the fall period. During spring, fourth-graders from other towns within a roughly fifty-mile radius are also brought in to spend most of a day on the prairie. As a result, over the years several thousand students have been able to see the prairie firsthand and, under the guidance of Deb Hauswald and other staff and volunteers, learn about both its natural history and human history.

There are also other seasonal educational opportunities for both children and adults at Spring Creek, such as participating in one-day "BioBlitz" surveys of plants and animals in late June or attending early fall festivals such as Twilight on the Tallgrass (2010) and Harvest of Traditions (2011). During the Prairie Festival in 2011, visitors could walk the Prairie Appreciation Trail and stop at sites where they could catch live grassland insects with nets and identify them, capture and examine aquatic invertebrates from one of the prairie wetlands, draw or color images of prairie plants and birds, or learn about some of Nebraska's endangered species.

During two days in late October 2011, I accompanied nearly 150 students from Lincoln's Roper Public School as they ventured out on the prairie, most of them probably having never before set foot on the grasslands that were so familiar and vital to earlier generations of Nebraskans. Divided into small groups, each with about a dozen students, their teacher, and a Spring Creek mentor, the kids were soon literally immersed in the tall grasses, some of which were twice as tall as them.

At an early stop the children learned about the complex grassland composition of a tallgrass prairie, with its hundreds of kinds of grasses and other plants, many of them wildflowers. They also heard about the prairie's interdependent populations of insects, which often exchange pollination for food in the form of nectar, pollen, or vegetation.

The students also learned of the poison juices of the common milkweed. These chemicals (alkaloids) protect it from being eaten by nearly all insects except the larvae of monarch butterflies and milkweed beetles, both of which have developed ways of neutralizing the plant's lethal effects. The children also enjoyed seeing the dispersal abilities of milkweeds, by releasing their parachute-like seeds into the air and watching them drift away in the breeze.

Another early stop was at the bottom of a long hill, marked by depressions that are the remnants of wagon ruts formed during the late 1800s. At that time wagon trains from Nebraska City cut through the then-virgin grasslands and passed directly over the middle of Spring Creek Prairie, carrying supplies to Fort Kearny, some 120 miles west along the central Platte River.

From the bottom of this hill the kids worked their way slowly upward, sometimes stopping to look at grasshoppers or other insects or to gaze at the seemingly limitless expanse of prairie all around them. About halfway up the hill they assembled on a grassy slope, where they were asked to sit, relax, and close their eyes. They then were told to imagine that the time was the 1860s and that each of them was a badger, living in a burrow and surviving amid the quiet of the prairie landscape. The kids were next asked to imagine hearing the sounds of an approaching wagon train for the first time, and how from that time onward their lives were forever changed, as were those of all other prairie inhabitants and, eventually, the prairies as well.

Later, small groups used nets to capture insects and, with the help of mentors, learned the basics of identifying them. Others used plastic hoops to mark a small circular patch of prairie and then closely examined the variety of plants that were found within it. Still others drew on blank postcards and then colored them using natural pigments squeezed from the berries of prairie plants.

After two hours on the prairie, the kids returned to the education building for picnic lunches. They then returned to the prairie for a brief afternoon session before taking a school bus back to Lincoln, tired but filled with memories likely to endure for a lifetime.

It is impossible to know if these unique experiences will permanently alter the children's perceptions of Nebraska's prairie heritage, but letters received later by Spring Creek provide some evidence. One of them reads, "Thank you for teaching me about the prairie. I learned about insects, different kinds of plants, and grasses. I liked drawing pictures with berries, flowers and leaves. I also enjoyed catching insects with a butterfly net and a jar. Now I want to go there again with my family, to see many insects and animals! Your friend, Hee Jo."

I hope that not only Hee Jo but also many of the other children will indeed return to the prairie again and again. It will help them to become emotionally attached to the natural world and to understand the value of preserving habitats such as Spring Creek Prairie. The prairie is an ever-changing and ever-fascinating tapestry of life, teaching important ecological lessons not so easily found in books, or so willingly assimilated. It is one of Nebraska's great natural jewels, which, like the Missouri and Platte Rivers, Chimney Rock and Scotts Bluff, provide visual reminders of who we as a state and nation are, and quite literally where we came from. It should also remind us of the importance of preserving and protecting all these great icons of historic America.

PART TWO

Seasonal Enchantments

Whooping crane

SPRING

Greater prairie-chicken

7

The Snow Geese of the Central Flyway

Walking down the hall of our university's biology building on a winter day during the late 1990s, I overheard two young men recounting their recent activities. One was proudly telling the other that he and some friends had killed over 120 "sky carp" the previous Saturday. Initially I had no notion of what he was talking about, but it soon became apparent that he was talking about snow geese and that he had exploited newly relaxed regulations that permitted winter snow goose hunting with few limits on the number of birds that could be shot in a day. I could scarcely imagine why anybody could take pleasure in killing that many geese, much less brag about it. Since childhood, snow geese have been my symbol of unmatched beauty and grace in the natural world.

On my birding trips to the Platte Valley later that spring and during following ones, I came to see all too clearly the effects of spring hunting on the behavior of both snow geese and other waterfowl. All the birds had abandoned their traditional spring stopover sites owing to the disturbance caused by hunting activities, and they were far more wary than before. Along marsh edges I often found dead hawks that probably had happened to stray too close to a goose blind; somehow, many waterfowl hunters still have the medieval idea that the only good hawk is a dead one. I felt ashamed that for a few years as a youngster I, too, had eagerly hunted ducks with my brother and father, until I was able to buy my first long-lens camera.

Since the late 1990s I have come to accept the idea of spring hunting of snow geese as one of the inescapable aspects of modern life, brought on by an ever-expanding human population, an associated increase in human violence, and a decreasing understanding of the natural world. I have also realized that my 1975 book *Song of the North Wind* describes an environment that was rapidly changing and that the three decades since its publication have brought about massive increases in continental snow goose populations and distributions. These changes have directly resulted from various human environmental manipulations, including an increase in available food crops on the wintering grounds and spring migration routes, an effective system of wildlife refuges that produces locally massive concentrations of geese, and the gradual climatic warming of arctic breeding grounds that has greatly improved breeding success rates in these previously highly marginal nesting environments.

As a result, I decided that I must revise my snow goose book. In the summer of 2009 I reviewed the technical literature of lesser snow geese as well as the other "light geese" of North America, namely the smaller Ross's goose of western Canada and the larger greater snow goose of eastern Canada's high arctic. Neither was a part of my original story, but for various reasons the fates of these birds have become increasingly intertwined with that of the lesser snow goose and need to be considered with it.

The Southampton Island snow goose population (in northern Hudson Bay) that I described for my 1975 book consisted of about 156,000 breeding birds in 1973. By 1979 it had grown to 233,000, and by 1997 it had reached 716,000. Other major Canadian snow goose colonies include Baffin Island, with 1.76 million breeding birds by 1997, the central Canadian Arctic, with 816,000 in 1998, and the western Canadian Arctic, with 580,000 by 2002. Including immature birds and other nonbreeders, the lesser snow goose population in the eastern Canadian Arctic had reached nearly four million birds by the late 1990s. There were also about one million lesser snow geese in the central Canadian Arctic by 1998 and 753,000 in the western Canadian and Alaskan Arctic by 2002. In addition, up to about 100,000 snow geese nesting on northeastern Siberia's Wrangel Island annually migrate through Alaska to winter in California's Central Valley. There were thus probably at least five million lesser snow geese alive at the start of the twenty-first century. The other light geese have also expanded their populations. The high-arctic and more easterly oriented greater snow goose population that mostly breeds farther north than the lesser snow goose and winters along the Atlantic coast first reached about one million birds by 2006. It had grown at an 8 percent annual rate since 1965 and attained an all-time high population estimate of 1.4 million by 2009. The Ross's goose population of northwestern Canada also reached or exceeded a million birds by 2001. These population changes occurred remarkably rapidly. In 1965–67, thirty-seven snow or Ross's goose colonies in the Queen Maud Gulf along the arctic coast of Canada's Northwest Territories had 44,300 nesting birds, with 77 percent of them Ross's geese. By 1988 there were fifty-seven colonies totaling 467,000 snow or Ross's geese, with about 60 percent of them lesser snow geese. That twenty-three-year period saw a 7.7 percent annual rate of population increase among Ross's geese and a 15.4 percent annual growth rate for lesser snow geese. Much of the snow goose's remarkable increase probably resulted from immigration out of colonies in the eastern Arctic, since a 15 percent rate of annual increase in geese is much higher than would be possible through local reproduction alone.

In recent decades, refuge management changes, altered agricultural practices, and milder winters have had major effects on midcontinental snow goose migration patterns, as to both timing and major wintering sites. Far more snow geese now winter in the Missouri Valley of Kansas and adjacent Missouri than was the case in the 1960s. Many of these birds are from the Southampton Island colonies.

Snow geese occur in both white and blue phases. During the 1950s the incidence of blue-phase birds in the Southampton Island colonies was 30–35 percent. Over the forty-year period from 1967 to 2006, blue-phase birds made up 27.5 percent of all the snow geese counted in the Great Plains region during Christmas Bird Counts. There is thus no evidence that either of these genetically based plumage types has shown a selective advantage over the other in correlation with changing arctic climates. And, although Ross's geese historically were entirely of the white-plumage phase, a few blue-phase adults have been found in recent years. This very rare plumage variant in Ross's geese, with a frequency estimated at no more than one in ten thousand birds, probably resulted from gene exchange during occasional hybridization with blue-phase snow geese.

American hunters were killing about a third of a million snow geese in the Mississippi and Central Flyways during regular hunting seasons of the late 1960s. At that time the midcontinent population of snow geese totaled about 1.5 million birds. Since 1972, when both flyways had daily shooting limits of only four snow geese, the continental light goose populations have all shown almost continuous proportional increases, and hunting regulations have been relaxed accordingly. In the Central Flyway states between North Dakota and Texas, daily bag limits were increased from five to seven birds in the 1980s and to ten birds in 1992. The waterfowl hunting season was also increased from 88 to 107 days, the maximum permitted by the Migratory Bird Treaty between the United States and Canada. Changes in this treaty during 1995 permitted the extension of the snow goose hunting season to March 10 in part of the Central Flyway, although in Nebraska the first spring season didn't take place until 1998. Other restrictions were also relaxed, including even-larger or unlimited daily bag limits, allowing shooting to continue until a half hour after sunset, the use of electronic calls to help entice geese into shotgun range, and permitting more than three shells in shotgun magazines.

By the mid- to late 1990s the midcontinent snow goose population had reached about three million birds. It was eventually decided by conservation agencies of both the United States and Canada that the lesser and Ross's goose populations should be reduced to half the late 1990s levels, and the greater snow goose population to five hundred thousand birds. Annual kills of these so-called "light" geese in the United States and Canada gradually increased from an average of 581,000 during the 1980s to more than a million between 1998 and 2002, as hunting regulations were relaxed and seasons extended. Special goose seasons between 1998–99 and 2001–02 thus helped increase the total continental light goose kill to an average of 1.3 million over the first four years of these "Conservation Order" hunting seasons.

These progressively increased opportunities for late winter and spring goose hunting were balanced by an increasing wariness of the geese and other compensatory factors, since total light goose kills leveled off and have declined somewhat after reaching a high point of 1.55 million during the 1999–2000 season. For the four-year period between 2005–06 and 2008–09, the combined U.S. and Canadian snow and Ross's goose kill averaged 890,000, or only about 60 percent of the record 1999 numbers. Considering the continued spectacular increases in light goose numbers in spite of this high level of hunter-caused mortality, the targeted population levels just mentioned for all three goose populations are unlikely to ever be reached through even more liberalized hunting regulations and expanded seasons. It also seems unlikely that these changes encouraging ever more lenient goose killing have engendered any new hunter understanding of and respect for their prey or for nature generally.

Snow geese populations in the Great Plains have not only increased tremendously in the past few decades, but many have also shifted their spring migration pattern from the Missouri River at least one hundred miles west into the central Platte Valley. This route change has brought more than a million snow geese into contact with several million sandhill cranes, cackling geese, Canada geese, and greater white-fronted geese. Snow geese and other geese usually arrive in the valley slightly earlier than do the cranes, and they are more prone to forage in the Rainwater Basin south of the Platte than to be concentrated like the cranes to the immediate Platte Valley. However, snow geese certainly compete directly with both cranes and other geese for corn throughout that region, as well as with an expanding deer population.

The geese and sandhill cranes have greatly benefited from the revolution in corn-growing technology in the Platte Valley, where production increases of about sixfold followed World War II. Greatly expanded irrigation, fertilization, and chemical pest management have all combined to produce a corn-growing bonanza in that region and have helped make Nebraska one of the major corn-growing states in the country. By the end of the twentieth century, record-setting annual corn crops were being grown statewide, with the Platte Valley responsible for nearly 40 percent of the state's total crop output. Between 1998 and 2003 the average annual state corn production was 1.1 billion bushels, and between 2004 and 2009 it averaged 1.4 billion bushels, or more than 150 bushels per acre. Assuming a harvesting efficiency of 90 percent, there would be about fifteen bushels per acre left in the field for wildlife to consume.

This food bonanza attracted a crowd. Although in the early 1970s an estimated 200,000 sandhill cranes staged there during spring, by the early 1980s crane numbers had increased to about 250,000, of which perhaps 200,000 were arctic-bound lessers. By the late 1990s maximum spring crane counts in the Platte Valley approached or exceeded 500,000 birds, although subsequently the population has apparently stabilized.

Evidence has been accumulating that, during their time in Nebraska, these midcontinent sandhill cranes are now unable to accumulate the levels of fat reserves that they had achieved in the 1960s and 1970s. This is probably the result of increased food competition among the millions of geese and cranes and a progressively improving corn-harvesting technology. Like the snow geese using the Platte Valley, the sandhill cranes are probably also now departing for their tundra breeding grounds in less-than-optimum breeding condition. But, like lesser snow geese, they have also benefited from the recently warming weather conditions in the Arctic during nesting and have continued to achieve successful reproduction.

Larger populations and increased hunting of midcontinent cranes since the 1960s have resulted in a recent annual hunting mortality (including those shot but not retrieved) approaching a record number of forty thousand birds, which represents most of this population's overall annual recruitment. The current near stability, if not decline, of the crane population makes it less likely that they will experience a population disaster such as that now facing snow geese. Because of the high level of use of the Platte Valley by endangered whooping cranes, sandhill crane hunting has never been allowed in Nebraska. For the same reason, spring goose hunting is also not allowed along a several-mile-long corridor on both sides of the river, which thus provides a safe feeding and roosting sanctuary for the cranes as well as other birds. The result is a spring bird spectacle perhaps now unmatched anywhere in the world, and one occurring only a few hundred yards off I-80! If I could have my wishes fulfilled, one would be that all Nebraskans might visit the central Platte Valley during March and, armed only with binoculars or a camera, absorb the sights and sounds of our state's most unique, most beautiful, and most fragile natural treasure.

8

A Congruence of Cranes

WITH KARINE GIL-WEIR

There are some people, sometimes referred to as craniacs, whose calendars recognize only two seasons: crane season and the rest of the year. In Nebraska the prime season for observing sandhill cranes occupies only about eight short weeks in spring, from mid-February (or whenever the Platte River fully thaws) to about mid-April. The usual peak crane populations occur during late March but rarely may be as late as early April. Sandhill crane numbers quickly trail off during the first warm days and with the south winds of April, when migratory conditions become optimum. Then the birds often depart almost en masse, leaving the state about the same time that the first whooping cranes are arriving from their localized wintering area (Aransas National Wildlife Refuge) in coastal Texas.

Research by Dr. Karine Gil-Weir indicates that the spring crane population of the central Platte Valley has been increasing very slowly during the past decade. The population peak during 2010 reached 479,000 sandhill cranes; however, from 1998 to 2010 the average was approximately 300,000 cranes, and the peak occurred most often during the fourth week of March. Spring arrivals have occurred in three chronologic "waves" over an eight-week period, perhaps in relation to the birds' relative breeding status or to the relative desirability of the various roosting sites. The earliest average arrivals and departures occurred in river stretches between U.S. Highway 281 (directly south of Grand Island) and Wood River, followed sequentially by the stretch from Wood River west to U.S.

Highways 34 and 10 and by the segment from Shelton west to Odessa. Roosts of up to seventy-two thousand birds were found; larger roosts were associated with long-term river management activities such as roost and channel rehabilitation. Departures occurred six to eight weeks after the river segments were initially occupied.

The sandhill cranes are present in Nebraska for an even shorter period during fall, mostly from late September to early November, peaking in early October. Relatively few Nebraskans see them then, as they tend to overfly the state, frequently passing directly from fall staging areas in western North Dakota and adjacent Canada south to temporary migratory stopover sites south of Nebraska. The most important of these southern migratory stopover sites are a few wildlife refuges in Kansas (especially Quivira National Wildlife Refuge) and Oklahoma (primarily Salt Plains National Wildlife Refuge), which are only a few hundred miles from the cranes' final winter destinations in Arizona, New Mexico, Texas, and northern Mexico. Their major Nebraska spring staging area, the Platte River Valley, is liberally sprinkled throughout the fall with trigger-happy waterfowl hunters, so that stopping there during autumn is a recipe for potential death, even among federally protected species such as cranes.

It is true that in recent years a very few sandhill cranes have remained in Nebraska throughout spring and summer, and some have even successfully bred here. These nestings have most often occurred on rather remote wetlands such as those of the Rainwater Basin and the Nebraska Sandhills, the latter having been part of their historic Great Plains nesting range during the nineteenth century. Cranes become extraordinarily secretive while breeding, so some of these occasional nestings may go undetected.

Bird-loving Nebraskans often speak proudly of the sandhill cranes as "our sandhill cranes," although the name "sandhill crane" was not historically derived from their occurrence in the Nebraska Sandhills. As a result, Nebraskans tend to underestimate or be unaware of the importance of other regions in the birds' annual cycle, which stretches at least four thousand miles over two continents, from northeastern Siberia, across Alaska and Canada, and south to northern Mexico. The birds are able to survive the rigors of breeding in arctic tundra, where freezing temperatures and blizzard-like conditions may occur at any time over the brief summer months. There predators such as golden eagles pose potential threats to adults, and arctic foxes, gulls, and jaegers are on constant alert for untended eggs or chicks. From late fall through winter the cranes must contrastingly adapt to semidesert or even desert conditions, where water and food supplies are limited and where at least one out of twenty is likely to be killed by hunters in the name of sport.

Between these two extremes, the cranes must traverse a distance up to almost twice as far as that from San Francisco to New York. Twice yearly they must travel over largely trackless lands of tundra, boreal forests, and grasslands, skirting the edges of North America's highest mountain ranges and visually navigating by the simple collective memories of the flock. Any single human able to complete this achievement in the face of all these collective dangers would be awarded a medal for personal fortitude and heroism. What the cranes receive instead is a barrage of gunfire over almost their entire route.

Many of the lesser sandhill cranes that migrate each fall through Nebraska begin their journey in Siberia, crossing the Bering Strait in late August and passing over the vast Yukon-Kuskowkwim delta of western Alaska, where they are joined by the local breeders, possibly in similar numbers. These birds move northeast along the Yukon Valley until they reach its confluence with the Tanana River, when they turn east and follow the Tanana southeast to a staging area upstream near Delta Junction. A few thousand of them stop for a few weeks at the edge of Fairbanks to rest and forage at Creamer's Field Migratory Waterfowl Refuge, where their arrival is eagerly awaited and celebrated with the annual Tanana Valley Sandhill Crane Festival. By early September tens of thousands of sandhill cranes move east along the Alaska Range and pass in front of the majestic north face of Mt. McKinley, North America's highest mountain. Except for the cranes breeding south of the Alaska Range, which migrate along the Pacific coast and winter in California, all of Alaska's sandhills then continue to fly southeastwardly. After leaving Alaska they follow the upper tributaries of the Yukon River between the Pelly and Selwyn Mountain ranges and continue southeast over Canada's boreal forest and onward into the northern Great Plains of Alberta and Saskatchewan. There, in spite of heavy local hunting, they can rest and forage for several weeks in grain fields, readying themselves for the long and dangerous trip over the central Great Plains, where they will face crane hunters in every state that they pass though except Nebraska.

Of all the nearly thirty thousand sandhill cranes killed legally each fall in midcontinental North America, the largest number are shot in Texas, followed closely by North Dakota. These killings destroy lifelong pair bonds and disrupt family bonds, probably making the young more vulnerable to hunting-related mortality. Given that the species does virtually no economic damage to humans and is regarded in many cultures as a symbol of peace, longevity, and fidelity, killing cranes seems better described as a sacrilege than as a sport.

By November large numbers of sandhill cranes have reached the southern Great Plains in Kansas. The main fall migration begins about October 8 and continues until late November. During the Audubon Christmas Bird Count of 2006–7, forty-eight thousand sandhill cranes were counted at Quivira National Wildlife Refuge, a national high-count record for that year. In Oklahoma the largest numbers gather at Salt Plains National Wildlife Refuge; about twenty-five thousand sandhill cranes were reported there during the Audubon Christmas Count of 2002–3. By December most sandhill cranes have continued farther south; in the 2008–9 Christmas Count thirty-three thousand sandhill cranes were observed at Muleshoe National Wildlife Refuge in northern Texas, and ninety-five thousand were seen there during the 2007–8 count. In the 2003–4 and 2004–5 counts the largest numbers (twenty-one thousand and seventeen thousand) were seen in southeastern Arizona, the westernmost wintering area used by the midcontinental population of lesser sandhill cranes.

The sandhill cranes wintering in southeastern Arizona are of special interest, as they are the southwesternmost wintering cranes that pass though Nebraska, staging each March in the North Platte Valley. They also are on average the smallest of the lesser sandhill cranes and probably migrate the farthest, as they nest in northeastern Siberia, roughly four thousand miles away. Six telemetry-equipped sandhills that wintered in New Mexico and Arizona left their wintering grounds in early to mid-March. Two of them went to breeding grounds on the Yukon-Kuskokwim delta, while the other four telemetry-equipped cranes traveled to Siberia's Chukotka Peninsula, arriving in mid- to late May, about the time the tundra becomes snow-free. Lesser sandhills are now known to breed as far west in Siberia as the Lena River delta.

For more than forty years a celebration of the cranes and other migratory birds has been held during mid-March in Nebraska, primarily through the planning and sponsorship of the Nebraska members of the National Audubon Society. Starting in 1970 and now known as the Audubon's Nebraska Crane Festival, it is one of the longest-running bird and wildlife celebrations in the country. Together in more recent years with the Rowe Sanctuary and Ian Nicolson Audubon Center near Gibbon, the celebration has done much to stimulate spring tourism in the central Platte Valley. It has also helped educate people from around the world about sandhill cranes and the other natural attractions of Nebraska's central Platte Valley.

As described in chapter 2, a thirteen-year Platte River Recovery Implementation Program began in 2006, jointly supported by agreements among the federal government, Colorado, Wyoming, and Nebraska. It is providing over $100 million for wetland purchase and restoration in the central Platte Valley to improve habitats for whooping cranes and three other nationally endangered and threatened species. Other major influences on the conservation of sandhill and whooping cranes and their habitats in the Platte Valley are several nonprofit conservation groups, such as the National Audubon Society, the National Wildlife Federation, the Nature Conservancy, and the Crane Trust.

Established in 1978, the Crane Trust (known as the Platte River Whooping Crane Habitat Maintenance Trust until 2012) was formed as part of an environmental settlement among the National Wildlife Federation, the State of Nebraska, and the Missouri Basin Power Project involving the downstream ecological effects of a dam being constructed on a North Platte River tributary in Wyoming. It has since played an important role in population monitoring and habitat restoration in the central Platte Valley, and in the past few years it has expanded its educational outreach by incorporating the Crane Trust Visitor Center at I-80 exit 305 into its purview. The Rowe Sanctuary and Ian Nicolson Audubon Center and the Crane Trust Visitor Center offer birders a wide choice of crane-viewing opportunities, ranging from single-person overnight blinds for professional photographers, situated only a few dozen yards from crane roosts, to blinds accommodating up to about thirty persons, ideal for groups and the general public.

Collectively, cooperation among state, federal, and many nonprofit conservation groups, as well as contemporary farming practices and strict control of the Platte's vital water resources, has allowed Nebraska to develop economically beneficial ecotourism from this largest concentration of cranes in the world. It represents one of the world's greatest migratory spectacles, easily equal to the great migrations of wildebeests across East Africa's Serengeti and Masai Mara plains or the caribou herd movements in arctic Canada and Alaska as they plod to and from their calving grounds. And, at least for Nebraskans, the spectacle is no more than a few hundred miles away!

9

The Whooping Cranes Are Still Surviving Tough Odds

WITH KARINE GIL-WEIR

Although many Nebraskans have had the indescribable pleasure of watching tens of thousands of sandhill cranes overhead, or even seeing them roosting on Platte River bars and islands during spring migration, only a tiny handful can say that they have ever seen whooping cranes in Nebraska. The sheer odds against it are daunting. Compared with 450,000–500,000 sandhill cranes migrating through the state each March, there are now less than 300 whooping cranes in the flock that annually migrates from Aransas National Wildlife Refuge, on Texas's Gulf Coast, to Wood Buffalo National Park, straddling the border of Alberta and Canada's Northwest Territories. In addition to to this numerical population disparity, whooping cranes migrate somewhat later in spring than the sandhills (during April in Nebraska), after most crane-watchers have gone home. During daytime foraging they also usually frequent rather remote wetlands far from any roads, and they generally move in small groups—a pair or a family or extended family that often consists of a pair and one or more generations of their offspring.

I have spent over fifty years studying, thinking, and writing about cranes and have observed cranes from Alaska to Arizona annually since 1961. My younger colleague, Dr. Karine Gil-Weir, engaged in full-time research on crane longevity, familial relationships, and migratory traditions in Texas and Nebraska from 2002 to 2011. Our devotion to these birds might be considered obsessive-compulsive behavior by many, but the sights and sounds of wild cranes are as intoxicating to us as the odors of tropical wildflowers. To be able to experience them only once would be as painful as being condemned to observe only a single sunset during one's lifetime.

A lifelong attraction to cranes among many people means that we now know as much about the lives of sandhill and whooping cranes as we do about almost any other North American bird. In 1941 only twenty-two whooping cranes existed in the wild (sixteen in Texas and six in coastal Louisiana). The Louisiana population was extirpated in 1949. It was not until 1954 that their Canadian breeding grounds were discovered (serendipitously, they are located in a remote subarctic area that had already been preserved, as Wood Buffalo National Park), and it was not until 1986 that the world population reached one hundred individuals. The whooping crane was listed as a nationally endangered species in 1972, which made available federal funding for research and for developing a survival strategy, although a wildlife refuge (Aransas National Wildlife Refuge) had been established to protect their wintering grounds almost a half century previously.

It is an ironic fact that the whooping crane's perilous population status has protected the Platte River from destruction by powerful irrigation interests, through the identification of the central Platte Valley as a critical habitat for the species. The threat of the extinction of the whooping crane, along with three other threatened or endangered birds and fish, has helped preserve Platte habitats for many other water-dependent species and set the stage for the enactment of a massive Platte conservation effort in 2006. This plan approved a multimillion-dollar, thirteen-year Platte River Recovery Implementation Program for wetland preservation and restoration in Nebraska, involving the federal government and the three states located in the Platte River drainage.

The central Platte Valley was the first region between Texas and the Canadian border to be federally designated as critical habitat for migrating whooping cranes. Other Great Plains sites that have since been similarly identified as critical habitats include Oklahoma's

Salt Plains National Wildlife Refuge and Cheyenne Bottoms State Wildlife Area and Quivira National Wildlife Refuge, both in Kansas. Among more than five hundred spring observations of whooping cranes obtained between 1943 and 1999 and analyzed by Jane Austin and Amy Richert, the largest percentage (69 percent) were from Nebraska locations, proving the critical importance of the Platte and other Nebraska rivers to migrating whooping cranes. There were progressively fewer migratory sightings in North Dakota, Kansas, South Dakota, Montana, Oklahoma, and Texas. Among nearly five hundred fall migration records, the largest share (33 percent) was also from Nebraska, followed sequentially by Kansas, North Dakota, Oklahoma, South Dakota, Montana, and Texas.

In Nebraska most historic sightings of migrating whooping cranes have occurred along the central Platte River, but they also have included the Niobrara, Middle Loup and North Loup rivers. In 2010 the most frequently visited counties during spring were Custer (nineteen records), Kearney (eight records), and Buffalo (eight records). Austin and Richert found that most locations where cranes have been observed were over a half mile from any human structures or developments. Most were more than a third of a mile from the nearest power lines, and about half of all the roost sites and two-thirds of the foraging sites had unobstructed visibility for more than a quarter mile and were associated with river widths greater than seven hundred feet. Clearly, visibility and distance from human activity are important aspects of whooping crane requirements. They also need access to wetlands for both foraging and nocturnal roosting. Like sandhill cranes, they prefer to roost in shallow water, well away from heavy shoreline or island vegetation.

Austin and Richert also determined that the average date for spring migration sightings is April 6 in Oklahoma, April 12 in Kansas and Nebraska, April 19 in South and North Dakota, and April 26 in Montana. In Nebraska, records that I have summarized similarly indicate that the peak spring migration period is the two-week-period of April 1–15 and the peak of fall migration is October 11–24. Starting in 1979, individual whooping cranes have been arriving in Nebraska much earlier than previously had been the case, probably because they had become socially attached to sandhill flocks that now typically arrive in mid-February rather than early March.

Partly in order to better document the role of individual cranes in migration and their population structure, a program of color-banding juveniles was undertaken in 1977 (when there were only sixty-two adults and thirteen juveniles at Wood Buffalo Park) and was continued until 1988. Gil-Weir has analyzed the survival and individual reproductive success from these banding efforts. Based on subsequent observations on the breeding grounds, on migration, and at Aransas, she determined that 24 of the 132 banded birds were still surviving as of 2009. All the survivors were at least twenty-two years old, and one had reached thirty-two years of age. Many cranes of various species are known to have survived more than thirty years in captivity (but rarely more than seventy), and some wild sandhills are also known to have had attained similar life spans.

Although first-year survival of banded birds averages 42 percent in Gil-Weir's analysis, survivorship of subadults increases annually until they reach four years of age. Older age-classes of whooping cranes have the highest mean survival rates and breeding success, reaching a maximum fecundity at about fourteen years of age. Indeed, one female at Wood Buffalo Park was still nesting at twenty-seven years of age.

Because whooping cranes are federally protected, various accidents are probably the major cause of adult mortality, especially collisions with overhead utility lines by migrating birds. By comparison, sport hunting is the cause of most mortality in sandhill cranes, resulting in the death of more than 5 percent of the lesser sandhill crane population annually. Thankfully, hunting for sandhill cranes in Nebraska has never been allowed, because of the special importance of the Platte to whooping cranes.

Partly because of their great longevity, and their strong family bonds, it is of interest to track individual and population characteristics among color-banded birds. Gil-Weir has determined such times of initial breeding, breeding success, length of pair bonds, incidence of remating after loss of a mate, and many interesting intergenerational associations. These include such information as the distances between the nests and wintering territories of parents and offspring and the common use of migratory stopover areas by close relatives, through four generations. Of special interest is the genetic legacy of a single identified pair that produced four descendants; from three of them that were banded it was possible to estimate that at least forty-three direct descendants contributed to the wintering Aransas population from 1977 to 2007. During that period the pair produced at least eighteen second-generation offspring that survived long enough to reach the Texas wintering grounds. The same genetic line subsequently produced seventeen third-generation birds and four fourth-generation offspring that likewise were able to survive the two-thousand-mile flight to Texas. Remarkably, many of these birds have continued to use the very same migratory stopover points as did their great-grandparents, showing the power of place memory in crane migration and the probable importance of migratory traditions in long-lived and long-distance migrants such as cranes.

The Aransas–Wood Buffalo flock has recently undergone some hard times, especially during the winter of 2008–09, when a severe drought in Texas resulted in the loss of 23 of its early winter population of 270 birds, a result of excessive freshwater withdrawals from the Guadeloupe and San Antonio Rivers by the local river authority that raised the estuary's salinity and killed much of the cranes' invertebrate food base, especially blue crabs. A year later, the total 2009–10 winter population at Aransas was about 263.

With the retirement of Tom Stehn, the refuge's whooping crane expert, who had carefully counted every last wintering crane annually for twenty-nine years, the protocol for counting the wintering birds changed, using incomplete sampling and extrapolated population estimates. The new survey method was first used for the 2011–12 flock, which produced an estimate of 254 birds in or near the Aransas refuge but had a margin of statistical error of 62 birds.

A more recent, 2012–13, estimate, of 273 birds (statistical range 178–362), suggests that the Aransas–Wood Buffalo crane population is simply holding its own. The overall world population of whooping cranes in early 2013 was about 600 birds, of which 445 existed in the wild, with the rest in confinement for breeding and research.

In the spring of 2010 nine chicks were banded and tagged with satellite and VHF radio-telemetry transmitters by the Crane Trust, the Canadian Wildlife Service, and other collaborators. This was the start of a four-year program that, from 2010 through the 2012 nesting season at Wood Buffalo Park, captured and tagged a total of thirty-one chicks out of a proposed total of forty. Since there had been a hiatus of more than twenty years since banding was last done on the breeding grounds, it is hoped that these current bandings will result in new sources of valuable population data for the understanding and conservation of this endangered species throughout its migratory corridor.

For the biologists who have worked so hard to restore whooping cranes, even small miracles such as the gaining of a few chicks per year must be rewarding. At times like these, with oil pollution in our precious Gulf Coast leading to the sight of majestic seabirds such as brown pelicans dripping with oil, the thought of immaculate whooping cranes flying high overhead is comforting, offering a reminder that we must all act to keep natural treasures such as cranes a reality.

10

The Strange Courtship of Prairie Grouse

At times I think that April is my saddest month. The last of the sandhill cranes are leaving the Platte Valley for northern breeding grounds. As I watch the last flocks disappear into the clouds like departing angels singing a final farewell, I am bereft, knowing that I will not see or hear them again for six months. My primary consolation lies in the fact that I know I will soon be hearing a different chorus, a sunrise serenade of grassland dancers, just as mysterious and beguiling as the departing cranes. But to hear and see this annual event requires more planning and even more patience than is needed for watching cranes.

At least for Lincoln or Omaha residents, one must leave by about 3:30 a.m. to drive the eighty-odd miles into southeastern Nebraska, where native prairie grasses still grow thick over glacial-molded hills. Then, in total darkness, one must find the right county roads and locate the best stopping point for setting up a blind. After setting up the blind in the dark, one must insert into it both oneself and all the necessary paraphernalia, such as a flashlight, binoculars, camera, spare lenses, tripod, gloves, a coffee-filled thermos, a stool, and perhaps a small tape recorder. Almost always the best place for a blind's location is atop a hill covered with low prairie grasses, at least several hundred yards away from tall trees or thick shrubbery and a quarter mile or more from any occupied dwellings. An advance scouting the day before, with critical odometer information recorded and the setting up of a few yellow flag markers to show the best predawn walking route, often makes the difference between finding the exact site and an entire morning's efforts wasted. Recent bird droppings and scattered grouse feathers provide the best clues to judging the center of mating activity.

If all goes well, one is settled in the blind at least a half hour or more before sunrise, before the eastern sky begins to brighten and the surrounding landscape features begin to take shape. If there is a full moon, an even earlier predawn arrival is needed, whereas a cloudy sky will mean that the curtain rising for the dawn serenade will be somewhat delayed. Then one must quietly wait, listening for early rising coyotes or perhaps the last great horned owl duet of the night. This is a time to be thankful for the preservation of these prairie relicts of the past—almost nowhere else in North America are there still countless locations where, without making reservations or paying a hefty viewing fee, one can watch and hear the dawn dance of the greater prairie-chicken.

I have called the greater prairie-chicken the spirit of the prairie; few other birds are so closely associated with native tallgrass prairies or are so sensitive to their destruction. It is a bird the color of autumn grasses, its feathers disruptively patterned in vertical stripes of switchgrass buff and Indiangrass brown, so that a motionless prairie-chicken simply fades into its background. Only its normally hidden under-tail coverts are conspicuously white. Also completely hidden beneath the elongated neck feathers of adult males are two patches of bright orange-red skin. Like secret signals, these areas are exposed only during the dawn and dusk mating ceremonies of prairie-chickens, when the males fill their throats with air, inflating the orange air sacs on each side of their throats and causing each side of their necks to resemble half tangerines. As these air sacs are expanded, the male utters a mellow and low-pitched cooing, something like that produced by blowing across the top of an empty bottle, but in a three-part cadence sounding to me something like *Old-Mul-Doooon*. Although this vocalization is soft, it can be heard for a mile or more under ideal conditions. The male also simultaneously stamps his feet rapidly, producing a soft drumming sound, and quickly fans and closes his tail feathers during each call sequence. While displaying, the male erects his long neck feathers so that a pair of ear-like appendages are formed on each side of his head and also tilts his tail vertically, exposing white under-tail feathers. This dramatic transformation of the bird's appearance, movements, and sounds produces a hypnotic effect on humans and, it would seem, on female prairie-chickens, for whom it is intended.

When the females arrive on the mating grounds, usually at about sunrise, they begin to inspect each male carefully, moving around the group like housewives searching for the best Thanksgiving turkey, but giving no outward indication of their possible preferences. The males in turn ratchet up the speed and intensity of their displays as each female approaches, and it is probably the relative vigor and perfection of an individual male's display behavior that helps females make their final mating choices. Not only are the males' minor display variations a possible basis for female choice, but of equal or greater importance is each male's relative position among the other males, a reflection of his ability to defend and maintain a desirable territory. Socially dominant and centrally positioned males ("master cocks") are often at least four years old and are the most effective at attracting and successfully mating with females. Indeed, even among a group of twenty or more interacting males, a single highly experienced and socially dominant male is likely to obtain at least 80 percent of all matings.

Many other open-country grouse, such as the North American sharp-tailed grouse, greater sage-grouse, and European black grouse, perform similar communal courtship ceremonies. These highly localized and strongly competitive congregations of display-ing males are called "leks," and their behavior is called "lekking," a Scandinavian word meaning to flirt. Lekking behavior appears to function biologically as a means of making certain that only the fittest males are able to attract mates and propagate the spe-cies' genetic line. Such a selective function requires an ability by females to rapidly and accurately assess all the males and likewise stimulates males to develop ever more effective ways of competing and attracting females. This process, called sexual selection, was first described by Charles Darwin and helps explain the evolution of such male traits as exhibiting conspicuous feathers or exposing colorful skin, uttering complex vocalizations, and performing extravagant postural displays from which females might select. It also accounts for the presence of such traits as antlers, horns, beards, and aggressive behavior among male mammals. Darwin realized that sexual selection may work reciprocally, with females detecting and choosing the most virile and strong males on the basis of such "secondary" sexual traits. Males likewise increasingly evolve traits making the fittest individuals able to outcompete other males, either through intimidation and physical dominance over others or by being more sexually attractive than other males. Over time, these interacting mating attributes produce ever more conspicuous sexual differences in adult behavior and appearance. Darwin hypothesized that our most refined human traits, such as having an aesthetic sense of beauty and a love of music, are ultimately attributable to sexual selection, and he even slyly inferred that women "first acquired musical powers in order to attract the opposite sex."

It is easy to forget about the theoretical basis for (and the human counterpart of) lekking behavior when watching prairie grouse and instead simply to become immersed in the action. One can detect spatial and behavioral differences among the males, as their territorial boundaries become apparent, and realize that some males are more self-assured, more aggressive, and more active than others. Thus it becomes easier to accept the idea that females are indeed able to choose desirable mating partners rapidly. Mating itself is brief and might be easily overlooked if one is not paying close attention. After a successful mating—and only a single mating is needed for a female to lay a clutch of twelve or more fertile eggs—the female leaves the lekking ground and begins to search for a nest site, which may be as far as a mile or so away. She will not interact again with males or other females until her brood is grown and autumn flock formation begins.

In Nebraska the males continue their daily display activities with diminishing enthusiasm until well into May, with some of the late matings probably the result of females having lost their original clutch and attempting a second nesting. The males play no roles in chick rearing or other familial duties. After a summer of molting and foraging, the older males usually return to the lek in early fall, apparently to reclaim possession of their spring territories or perhaps try to expand into space made available by the deaths of others. This fall display activity also attracts the attention of young males, who may become peripheral viewers or even minor participants. As each male grows older and more experienced, he is likely to move his territory ever closer to the middle of the lek, with the potential of eventually becoming a master cock if he lives long enough.

There are several possible options for visiting a prairie-chicken lek. Prairie grouse are probably most common in the eastern and central Sandhills, where optimum survival conditions are provided by a combination of native Sandhills prairie and access to corn and other crops that supplement winter foods. Public-access leks with ready-made blinds are available on a first-come, first-served basis at the Bessey Ranger District of Nebraska National Forest in Thomas County and at Valentine National Wildlife Refuge in Cherry County.

Commercial operations that offer views of prairie-chicken and sharp-tailed grouse leks in the relative comfort of permanent or school bus blinds are provided by the Switzer family at Calamus Outfitters, near Burwell, and by Mitch Glidden at the Sandhills Motel in Mullen. In southeastern Nebraska, the Big Blue Ranch and Lodge, located three miles west and three miles south of the village of Burchard and operated by Scott and Billie Kay Bodie, offers spring trips to a prairie-chicken lek as part of a lodging package.

No Nebraskan should consider his or her life complete without experiencing these drums of April. Like watching a star-filled Sandhills sky, seeing sandhill cranes in formation above the Platte, or canoeing the Niobrara River, it is a defining experience of life on the Nebraska prairie.

11

The Secretive Shorebirds and
Their Amazing Migrations

Certain ecological and landscape attributes of the state of Nebraska have placed it at the center of a vast, invisible aerial pathway known to biologists as the Central Flyway. To the west the Rocky Mountains provide the western boundary of a broad north-south corridor formed by the Great Plains, while to the east the Missouri River Valley offers similarly conspicuous landmark guidelines for north- or southbound birds. Nebraska's Platte River and thousands of other mostly small Nebraska wetlands are situated roughly halfway between the Gulf Coast and the transition zone between the northern Great Plains and the vast Canadian coniferous forests, the last geographic barrier to arctic-bound breeders. The Platte Valley thus represents the geographic "waistline" of a north-south pathway that is an important oasis of rest and foraging in the middle of one of the great avian migratory routes of North America, especially for cranes, waterfowl, and shorebirds. Because of its unique geographic location and abundant wetlands, the central Platte Valley and nearby regions attract countless numbers of nearly one hundred migrating wetland species every spring, most famously waterfowl and sandhill cranes.

The spring migration of geese and ducks through Nebraska is well documented; it involves nearly ten million birds and includes nearly thirty species. The migration is centered in the central Platte Valley and is concentrated during a few weeks in March, usually peaking a week or two earlier than that of sandhill cranes. Also like cranes, the constant overhead calls of Canada, cackling, greater white-fronted, Ross's, and snow geese can't easily be ignored, and the endless streams of shifting flight patterns projected against the sky aren't likely to be overlooked by even casual observers.

When I arrived in Nebraska in 1961, the astonishing spring migration of a half million sandhill cranes to the central Platte Valley was a phenomenon then known only to the local residents. I luckily learned of it during that first year and, like the cranes, have since been irresistibly drawn back to the Platte every year. Only now, a half century later and after I have written several books describing the sandhill cranes and their intimate connections with the Platte Valley, has this spectacle has begun to receive the worldwide attention that I believe it deserves.

Much less apparent and well known is Nebraska's role in the spring and fall migrations of shorebirds such as sandpipers, plovers, curlews, and godwits. These migrations are inconspicuous, in part because of the fact that most long-distance shorebird flights occur at night. Shorebirds also never migrate in the enormous flocks that are so typical of geese and cranes, and their flights are usually unaccompanied by loud calls. Yet they are massive if nearly invisible migrations, involving over thirty species and an estimated two hundred thousand to three hundred thousand birds. Many of the species travel from South American wintering grounds, and about half of them are bound for arctic tundra breeding grounds in Canada and Alaska. While resting and feeding between flights, the birds scatter across Nebraska's smaller and shallower wetlands, from the undulating loess plains of Rainwater Basin between the Platte and Republican Rivers, across the wet meadows of the Platte Valley, to the thousands of remote Sandhills wetlands nestled between the Platte and Niobrara Rivers.

Nebraska's wetlands are a small but important part of an intercontinental series of important stepping-stones used by shorebirds on their migrations. Some locations are especially valuable for certain species but may be little used by others. For example, the otherwise undistinguished agricultural fields and wetlands centered in Seward and Fillmore Counties are probably the buff-breasted sandpiper's single most important spring stopover area between its Argentine wintering sites and its high-arctic nesting grounds. Not far to the south, the shallow marshes of Cheyenne Bottoms Wildlife Area in central Kansas support nearly half of North America's total shorebird population, including more than 90 percent of the white-rumped, Baird's, and stilt sandpipers, long-billed dowitchers, and Wilson's phalaropes surveyed during spring in central and eastern North America. To the north the glaciated potholes and sloughs of the Dakotas and southern Canada provide the major stopping points on the spring flights of many northern prairie or arctic-nesting species.

The migration pathway of each shorebird species is mostly dependent on the distribution of crucial stopover wetlands, of which the shallow, clay-bottom wetland playas (locally called "lagoons") in the Rainwater Basin are particularly important. There the birds rest for a few days, foraging continuously during the daylight hours on invertebrates that they find by pecking, probing, and sieving actions, depending on the species' bill shape and length. Short- and stout-billed plovers such as killdeers typically run and peck at surface foods, using their large eyes and keen vision, while species with progressively longer bills, such as sandpipers, dowitchers, and snipes, typically probe in shallow water and mud, using only their sensitive bill tips to detect food. Phalaropes swim about on the surface in dizzying circles, using the turbulence thus produced to bring to the surface tiny edible items that can be easily grasped with their delicate bills. The long bills of curlews are bent gracefully downward, facilitating efficient probing in sand while keeping the head nearly horizontal; the bill of the female is noticeably longer and more decurved than the male's. The upturned bill of avocets is adapted to making scythe-like sweeps across the water surface, the upward curvature of the bill allowing for a maximum surface area of water to be intercepted. This remarkable diversity in bill shapes, bill lengths, and foraging behaviors is testimony to the powers of natural selection in fitting many potentially competing species into a maximum number of nonoverlapping foraging adaptations.

Collectively, the spring shorebird migration extends from late February or early March, with the arrival of killdeers, to early June, when the last of the arctic-bound species finally depart. Based on studies by Joel Jorgensen in the eastern Rainwater Basin, the peak of the shorebird migration there occurred in the second week of May (in eight of thirty-three species) during his 1997–2001 study. Based on a half century (the 1930s to 1980s) of migration reports from across Nebraska by members of the Nebraska Ornithologists' Union, I determined that ten out of thirty-four shorebird species exhibited a migration peak (median arrival dates) during the first week of May. Ten more exhibited peaks during the second week of May, indicating that the first half of May is the best time for finding shorebirds. By early to mid-May Nebraska's wetlands are warm enough that invertebrate life is abundant, and the birds are able to store sufficient energy to carry them still farther north.

Not all the shorebirds arriving in Nebraska have still farther to go. Many of the killdeers arriving about the middle of March quickly scatter across all the state's meadows and grasslands, sometimes even nesting in the parks or golf courses of larger towns. Killdeers are easily the most recognized of our shorebirds, and those nesting here are likely to have wintered in Central America, or even as far south as Venezuela. American avocets and black-necked stilts similarly arrive in western Nebraska during late April or early May, from coastal lowlands of Mexico in the case of the avocets. The stilts might have traveled from as far south as South America, where this species is widespread. Wilson's phalaropes also overwinter on inland wetlands from Mexico southward, but they are late migrants. Huge flocks of phalaropes migrate through the more alkaline wetlands of the Sandhills during early May, where many stop to nest. Avocets and stilts also prefer alkaline wetlands for breeding, so all are likely to be found using the same ponds, which are often rich in brine fly larvae, brine shrimp, and other salt-tolerant invertebrates.

Ruddy turnstones, sanderlings, and red knots, all fairly rare spring migrants, stop here only briefly on their headlong rush northward to nest on high-arctic tundras as far away as northern Ellesmere Island, the northernmost land mass in North America. Of these, the sanderling winters the farthest south, often traveling to southern Chile and Argentina, and sometimes even to Tierra del Fuego, roughly ten thousand miles from its northernmost nesting grounds. All of this flying and navigational power is packed into a bird weighing only slightly more than two ounces!

Some of our migrant shorebirds are more common in spring than fall, or vice versa. For example, the white-rumped sandpiper is one of Nebraska's most abundant spring sandpipers, but during fall these birds swing east to the Atlantic coast and are very rare in Nebraska. Similar, somewhat elliptical migrations occur in the American golden-plover, Hudsonian godwit, sanderling, and semipalmated sandpiper. However, the western sandpiper's major spring migration route occurs along the Pacific coast, and the birds are likely to migrate through the Great Plains in large numbers only during their fall migration.

Persons wanting to see the spring shorebird migration in Nebraska have many choices. In the western Rainwater Basin the broad and shallow marshlands of Funk Wildlife Management Area (two miles north of Funk) often attract great numbers of shorebirds, as do several other sites. Joel Jorgensen reported that he saw the largest numbers of spring shorebirds in the eastern Rainwater Basin at three federally owned waterfowl production areas: Harvard (three miles west of Harvard), Mallard Haven (two miles northwest of Shickley), and Massie (2.5 miles southeast of Clay Center). All of these are playa wetlands, which soon become entirely dry during years of low winter precipitation, whereas during very wet years the roads leading to the wetlands may be so soft as to make them impassable.

In the western Sandhills, Crescent Lake National Wildlife Refuge (twenty-five miles north of Oshkosh) is a prime place for shorebird watching. At times thousands of Wilson's phalaropes and American avocets may be seen on some of the refuge's more alkaline wetlands, such as Border Lake. In the northern Sandhills, Valentine National Wildlife Refuge, twenty miles south of Valentine, probably offers the best regional viewing opportunities. Least and Baird's sandpipers are notably common there during spring.

Of all Nebraska's commonly nesting shorebirds, the species that perhaps migrates the farthest from its wintering areas is the upland sandpiper. It typically arrives in early May after wintering in the pampas grasslands of Argentina, nearly five thousand miles away. In a sense it lives in a world of constant spring and summer, leaving the pampas at the end of each southern hemisphere summer to be greeted upon its arrival in Nebraska by yet another beautiful spring. Like the equally welcome calls of the long-billed curlew, the upland sandpiper's territorial and courtship calls provide Nebraska ranchers with tangible proof that each long winter is finally over.

These and other Nebraska shorebirds are fragile treasures whose destinies largely depend upon the preservation of wetlands that are scattered from northern Canada to southern South America. An important link in this pathway, Nebraska is one of a few places left in America where a person can stand, surrounded by native grasses and graced above by an unobstructed blue sky, while the distant whistles and bubbling calls of an upland sandpiper send shivers down the spine or a long-billed curlew can be seen standing majestically silhouetted against a grassy horizon. These are indeed living treasures worth protecting, especially by conserving our remaining wetlands.

SUMMER

Upland sandpiper and prairie phlox

12

Birds of the Tallgrass Prairie

Nebraskans may be justifiably proud of our tallgrass prairies; few other states have a larger number of protected tallgrass prairies that are open to the public for our enjoyment and educational opportunities. They include prairies located in state parks, such as Rock Creek Station State Park in Jefferson County, and state wildlife management areas (WMAs), such as the 1,120-acre Pawnee WMA, Pawnee County, and 600 grassland acres in Twin Lakes WMA, Lancaster County. There are also some federally owned restored prairies, such as at Homestead National Monument in Gage County and Boyer Chute National Wildlife Refuge in Douglas County. City-owned prairies include both virgin and restored prairies in Lincoln's Pioneers Park, and there are several owned by the Nature Conservancy. The University of Nebraska owns the historically famous 240-acre Nine-Mile Prairie, located nine miles northwest of Lincoln. The best-preserved and one of the best-studied tallgrass prairies in Nebraska is National Audubon's 800-acre Spring Creek Audubon Prairie. Located in the glacial moraine hills of southern Lancaster County, it was acquired by the National Audubon Society in 1999. It contains over 350 species of plants and is an education center as well as a pristine prairie (see chapter 6). Most of these sites are freely open to the public.

The adjacent states of South Dakota, Iowa, and Kansas also have many prairie preserves. In South Dakota, the Nature Conservancy owns about ten thousand acres of mostly tallgrass prairies at eight locations in eastern South Dakota, of which the seventy-five-hundred-acre Samuel H. Ordway Jr. Memorial Preserve in the tallgrass–mixed-grass transition zone is the largest. Iowa's few remaining tallgrass prairies are nearly all located in the loess hills region of western Iowa. The largest of these is the three-thousand-acre Broken Kettle Grasslands near Sioux City, owned by the Nature Conservancy. In the Flint Hills region of eastern Kansas, the National Park Service and the National Park Trust jointly manage the Tallgrass Prairie National Preserve, of eleven thousand acres, and the eighty-six-hundred-acre Konza Prairie is similarly jointly managed by the Nature Conservancy and Kansas State University.

Although dominated by a relatively few species of tall perennial grasses, tallgrass prairies are notable for the very large diversity of broad-leaved herbaceous plants that are also present, frequently comprising three hundred or more species. Most of these plants, technically classified as "forbs," fall into the general category of wildflowers, and it is the spectacular summer and fall array of native wildflowers that stimulates most people to visit prairies. For the discerning eye, not even the autumnal colors of our eastern deciduous forests can match the summer reds and pinks of prairie phloxes and roses, the later violet-purples of gayfeathers and coneflowers, the Indian summer yellows of goldenrods and sunflowers, or the coppery tones of late autumn grasses. It is this diverse plant life, whose seeds, leaves, and flowering parts support an interdependent population of insects and other invertebrates, that is the biological basis of the prairie's native bird populations. Of the roughly thirty species of North American birds that are ecologically closely associated with grasslands, the largest single component is made up of sparrows and other seedeaters, although even sparrows depend heavily on insects for proteins while feeding their young.

It is of interest to compare the species compositions of the grassland bird populations at Spring Creek Audubon Prairie and Konza Prairie, as both are relatively large grassland ecosystems with similar climates and vegetational attributes. Native species that breed in both locations and are particularly grassland dependent are the greater prairie-chicken; upland sandpiper; northern harrier; horned lark; field, lark, grasshopper, and Henslow's sparrows; dickcissel; bobolink; and eastern and western meadowlarks. Other birds breeding in both prairies but having somewhat broader ecological niches are the northern bobwhite, American kestrel, killdeer, cliff swallow, barn swallow, sedge wren, common yellowthroat, red-winged blackbird, and brown-headed cowbird.

Prairie birds are something akin to prairie grasses; they are often confusingly similar and usually not very colorful. Yet what they might lack in color they often make up for in both song and behavior. Few Nebraskans would admit to not being touched by the spring songs of meadowlarks, whose clear and melodic voices by late February often proclaim the end of winter long before it is admitted by any of the people who accept the calendar's insistence that spring always begins on the twenty-first of March.

Following the sometimes overly optimistic weather assessments of the meadowlarks, mid-March brings the next uniquely prairie concert—the dawn chorus of greater prairie-chickens. Displaying from distant hilltops as the grassy slopes are just being touched and illuminated by the rising sun, this rhythmic melody is so haunting and powerful as to make one almost believe in the presence of unseen spirits. It penetrates to some remote and perhaps ancestral part of the brain, producing a kind of restful mantra that makes one believe that all is well with the world, in spite of the overwhelming evidence to the contrary.

The complex mating choreographies of prairie-chickens extend into and through April, by which time most of the other prairie birds will have arrived. Red-winged blackbirds begin to utter their familiar *kong-kor-eee* notes from nearby willow-lined wetlands, simultaneously exposing their blood-red wing-covert epaulets to capture the attention of any nearby female. Male bobolinks establish territories in wet meadows, periodically releasing a cascade of song as they launch into the air and patrol their self-proclaimed properties. Grasshopper sparrows and Henslow's sparrows more quietly take up their territorial position in the previous year's tall grasses, uttering soft, insect-like songs that, together with their camouflaged dead-grass plumages, perhaps help protect them from aerial predators. One of the last of the major prairie songbirds to arrive is the dickcissel, which has to make a spring migration of more than a thousand miles from South America, where it winters from Colombia to French Guyana. Arriving in mid-May, the males in spring resemble a miniature meadowlark, with a yellow breast crossed by a black blaze, revealing their ancient blackbird-like evolutionary connections. Like the grasshopper and Henslow's sparrows, dickcissels have declined greatly across North America as their prairie habitats have disappeared, but they maintain a tenacious hold in eastern Nebraska, where a combination of weedy old-field habitats, Conservation Reserve Program grasslands, and tallgrass prairie relict stands provide ideal habitats. The dickcissel is the most common prairie breeding songbird species at Konza Prairie in eastern Kansas, and the same is true in eastern Nebraska sites such as Spring Creek Prairie. The breeding range of the dickcissel fairly closely corresponds with the historic parts of the cowbird's range, which once largely was associated with that of the bison but has since expanded to include much of the deforested pastures and croplands of North America.

Perhaps because of their great abundance, dickcissels are the favorite target of the brood-parasitic brown-headed cowbird in eastern Nebraska and are also prime victims at Konza Prairie. In Nebraska it is uncommon to find a dickcissel nest that does not have at least one cowbird egg among the host's eggs. In a Kansas study John Zimmerman found that, among 544 nests, only 46 percent were not parasitized, while the remainder had up to as many as at least six cowbird eggs! Since a female cowbird will lay only one egg in each nest she parasitizes, the local cowbird population must be remarkably high in Kansas prairies, and the same appears to be true in Nebraska. In other Kansas grassland studies parasitism rates on dickcissels have been found to range from 50 to 91 percent of all nests found, as compared with 22 to 50 percent of grasshopper sparrow nests and 70 percent of eastern meadowlark nests.

In spite of the large number of eggs that they lay, cowbird reproduction is not very efficient. John Zimmerman reported that only 7 percent of 132 cowbird eggs in dickcissel nests resulted in fledged young, although other Kansas studies suggest that a more typical fledging success rate is about 23 percent. However, he also found that nonparasitized dickcissel nests produced an average of 3.7 fledged dickcissels per nest in prairie habitats, whereas among parasitized nests the dickcissels fledged only 1.8 of their own young per nest. In larger bird species, such as the red-winged blackbird, competition for food by nestlings is probably more evenly matched, but highly effective begging behavior by the cowbird chicks may allow them to grow faster and be more likely to fledge than their blackbird nest mates.

In contrast to more famous nest parasites like the famous Old World cuckoo, cowbirds have not evolved highly specialized traits, such as mimicking the host's egg color and pattern or getting rid of the host young by pushing them or their eggs out of the nest, Instead, cowbirds rather indiscriminately drop their eggs into virtually every nest that they find, regardless of the host species. Some host species can recognize the alien egg and eject or otherwise eliminate it, while others simply accept it as one of their own.

In many songbird nests the young cowbirds often hatch slightly sooner than the host's young and immediately begin to obtain much of the food brought by the parents, simply by their more prolonged and more insistent begging. Although a dickcissel brood might survive the presence of a single cowbird chick in the nest, every additional cowbird increases the chances that most or all the dickcissel young will starve to death before fledging. Only after late July, when the cowbirds finally stop laying eggs, do the dangers from them subside. It is likely that a single female cowbird may lay fifty or more eggs in a single breeding season, and one female was proven to lay sixty-seven eggs in as many days!

In spite of the massive damage cowbirds have brought to grassland and forest-edge bird populations, the overall range of cowbirds has now stabilized, and their national population density has been stable or declining slowly since the mid-1960s. Probably the present greatest current threat to prairie bird populations is the continuing loss of grassland habitats through ever more conversions to cropland. Grassland birds have many other enemies, such as snakes, weasels, and even ground squirrels, which devour eggs and chicks. Avian predators include great horned owls, taking larger species such as quails and prairie-chickens, and fast-flying Cooper's and sharp-shinned hawks, which concentrate on progressively smaller birds. Natural disasters such as wind, snow, rain, and hailstorms all contribute to avian mortality, so that at times it seems impossible that any birds will survive. Yet year after year they appear on schedule, go through all the dangers and difficulties of life, and annually not only provide us with the joys of seeing and hearing them but also offer models of parental care and devotion. Even cowbirds have to be respected for their remarkable adaptive capacities to survive and reproduce in an unforgiving world.

13

Nebraska's City-Dwelling Peregrines

During the 1930s and 1940s there were about 500 pairs of peregrines in the United States, including about 210 active nests in the eastern states and 250–350 nests in the western United States. Additionally, the arctic peregrine, a tundra-nesting race, consisted of perhaps 150 pairs that nested in high latitudes from Alaska to Greenland. Over all of their range the birds typically nested on steep cliff sides, and because of this need for tall-cliff nesting sites, there are no firm historical nest records for Nebraska, although some evidence exists for possible breeding having occurred near Fort Robinson in 1903.

By 1960 the peregrine's North American population had crashed, mostly as a result of the effects of pesticides introduced after World War II, especially DDT. DDT enters the falcon's system through its consumption of poisoned prey, especially insect-eating birds, and interferes with the hormones regulating a female peregrine's reproductive system. As a result, she lays soft-shelled eggs that might be crushed by the incubating bird or that otherwise fail to hatch. The peregrine falcon population had reached critically low numbers in North America by 1970, when it was listed as nationally endangered. It was among the species to be added to the initial list of nationally endangered species when the Endangered Species Act was enacted in 1973, thus becoming eligible for federal research and population recovery efforts.

It was not uncommon during the late 1960s to walk across the University of Nebraska's Lincoln campus and see dead and dying songbirds after DDT had been sprayed to try to control summer mosquito populations. The publication of Rachel Carson's book *Silent Spring* in 1962 had alerted the country to the dangers of using such pesticides, but it took another decade of political and legal fighting before the sale and use of DDT in the United States was finally prohibited. In the late 1960s I published a community columnist piece in the *Lincoln Journal*, pointing out the health perils to both humans and wildlife of using DDT. I very soon received an irate letter from one of the vice presidents of Vesicol, the largest American producer of DDT and other agricultural chemicals, telling me that the substance was completely harmless, and that to prove it he sprinkled a teaspoonful of DDT on his breakfast cereal every morning! I didn't reply but have often wondered how much longer he survived.

The increasing construction of high-rise buildings in many U.S. cities has had an unexpected benefit for peregrines. Evidently accepting the idea that skyscrapers are nothing but artificial cliffs, peregrines began nesting on their highest ledges as their numbers slowly began to increase in the late 1970s. Cities not only provide such potential nest sites but also an abundant supply of prey in the form of urban birds, such as rock pigeons and starlings, and freedom from predators of newly fledged birds, such as great horned owls and golden eagles. Buildings that are at least ten stories high are selected; such heights are apparently the lowest that the birds will accept. Nests located at lower heights are perhaps too low for a newly fledged peregrine to launch from and gain enough flight speed to achieve aerial maneuverability control without crashing to earth.

By the late 1980s a major restoration program was underway for the peregrines, largely planned and funded by the Peregrine Fund, a nonprofit group centered in Boise, Idaho. This program involved monitoring and protecting active aeries in the western mountain states and reestablishing nesting birds in both western and eastern states by captive breeding and release ("hacking") programs. By the early 1990s over three thousand captive-bred peregrines had been released in the United States, and there were about one hundred active breeding pairs in the East, about twenty in the Midwest, and about four hundred in the West. The peregrine's population has since continued to improve, and it was removed from the nationally endangered list in 1999.

In Nebraska, efforts to establish peregrines in the state began in 1998, when seven juvenile birds were hacked at Woodmen Tower, then the tallest building in Omaha. Of these, five fledged, but none returned the following year. In 1989 five more falcons were hacked, including two males named Woody and Sky King.

In 1992 Woody returned to the tower and mated with Windy, a female that had hatched during 1990 in Des Moines, Iowa. The pair produced three young, Ariel, Zenith, and Skywalker, becoming the first peregrines known to have been produced by a wild pair in Nebraska. During the following year, Windy produced three infertile eggs.

In 1993 Woody's brood-mate Sky King returned to Omaha and mated to a female (Kay Cee) that had hatched the previous year in Kansas City, Missouri. In 1994 they produced a single fledged chick, Sokol, and in 1995 another chick was produced, but it died before being named.

In 1996 Zeus began his amazingly long reign in Omaha. He had hatched in Rochester, New York, during 1994. In 1996 he and his mate, Minnie (who had hatched in Winnipeg, Manitoba, in 1994), nested on the Woodmen Tower, producing four chicks and fledging two of them. In 1997 four more chicks hatched, but all died before fledging. In 1998 five young were fledged, and in 1999 four more were successfully reared.

By 2001 Zeus had acquired a new mate, Amelia, hatched in 1999 in Cedar Falls, Iowa. He had begun courting her the previous year, when she was still too young to breed. The new pair successfully hatched four chicks in 2001, but none survived to fledging. However, during the next three summers they fledged a total of eight young birds, including at least two every year.

Zeus acquired his third mate in 2006, a female (Hera) of unknown age and origin. In 2006 they managed to fledge four youngsters, followed by four more in 2007. The 2008 brood of two died before fledging, but in 2009 a male (Magnus) and three females (Inina, Willow, and Isis) were fledged, and in 2010 two males (Ponca and Otoe) and two females (Kioa and Dakota) were reared. By then about fifty peregrines had been successfully reared on the Woodmen Tower. In 2011 (the last year of Zues's reign), the Woodmen pair generated five females (Mae, Amelia, Sally, Rosa, and Melba)! In 2012 the Woodmen site (with a new male named Mintaka that had hatched two years previously in Lincoln) and Hera reared two females (Cass and Harney) and three males (Farnum, Dodge, and Douglas). In 2013 the same pair reared three females (Joslyn, Orpheum, and Durham), and one male (Big O).

The story for Lincoln is similar. I had long believed that Lincoln's capitol might support breeding falcons and had once suggested to city officials that a hacking program should be initiated. The question of hacking costs and the problem of interfering with the external appearance of the building made that proposal unsuccessful. However, during the summers of 1990–93 a male peregrine frequented the capitol, acting territorial and courting females. In 1991 the Nebraska Game and Parks Commission installed a nesting platform on the eighteenth floor, just below the capitol dome. Several females were seen around the capitol over several years between 1992 and 1997, but no nesting attempts occurred. The nest box was removed from 1998 to 2002 during the capitol's renovation but was reinstalled in March 2003.

By May 2003 a pair of peregrines had taken territorial possession of the capitol. The male (band 19-K) had hatched in Des Moines, and the female (Angel) had originated in Minneapolis. The pair produced two eggs that year, but neither hatched. During the following year a pair was also present, but no eggs were produced. After a protective roof was added to the nest box and a television monitor was installed in February 2005, the box quickly attracted a pair of peregrines. The male was the same one that had been present in 2003, but the female (A/*Y) was a new mate that had hatched in Winnipeg. The male chick that they fledged that year was appropriately named Pioneer. Pioneer was not only the first of the capitol's fledged peregrines but also a testimony to the efforts of the late John Dinan, the Nebraska Game and Parks biologist who had worked so hard to make the project a success.

In 2006 the same pair produced three fledged chicks, Willa, Bess, and Sterling, named for Willa Cather, Bess Streeter Aldrich, and J. Sterling Morton. In 2007 they successfully raised four, Boreas, Notus, Eurus, and Zephyrus, named after the four winds of Greek mythology. In 2008 four eggs were laid, but the nest failed. In 2009, with a newly designed nest box installed, the pair raised four more chicks, named after Nebraska rivers, the males Platte and Calamus and the females Nemaha and Niobrara. The chicks produced by the capitol pair have annually been given names on the basis of an annual free and open contest, in which anybody may suggest possible names.

Most recently, in 2010, four eggs were laid, of which three (two males and a female) hatched and fledged. In the annual chick-naming contest, they were given highly appropriate names for aerial hunters: Alnitak, Alnilam, and Mintaka, after the three bright stars in the belt of the hunter constellation Orion. In 2011 the capitol pair produced one male (named Lincoln). In 2012 the Lincoln pair produced two young, who were named Lewis and Clark, and in 2013 three chicks (two females, one male) were reared in Lincoln.

In recent years the Woodmen and capitol falcons have become tourist attractions, and it is possible to see the daily activities of both pairs during the nesting season through monitors in the two buildings. They may also be seen via Internet connections from the Woodmen of the World building (http://www.woodmen.org /falcon_cam.cmf) or the website of the Nebraska Game and Parks Commission (http://www.outdoornebraska.ne.gov/wildlife/webcam /peregrine/1).

Besides the convenience of seeing the family on television, one may watch their activities from the capitol grounds with the aid of binoculars or get very limited views from the observation deck on the capitol's fourteenth floor. Throughout spring and summer the adults engage in spectacular flights as they hunt for birds to feed themselves and their young. Often the prey is rock pigeons that are common in the city, but the falcons have been found to take a wide variety of songbirds. There are even accounts of the birds using the lights that illuminate the capitol after dark to attack nocturnal birds such as nighthawks.

During their chases, and especially their vertical dives, peregrines attain air speeds far in excess of any other species. By taking trained peregrines into airplanes and skydiving with them, it has been possible to make accurate estimates of peregrine air speeds during their full-out dives. In two such experiments, speeds of 183 and 242 miles per hour were calculated! This blinding-fast speed means that nothing below them has a chance of out-flying a peregrine. However, it also must require incredibly fine flight control and skill at that speed to home in on a bird far below that may be flying up to fifty miles per hour at a nearly right angle to the plummeting falcon and that is frantically trying to avoid being captured.

It seems especially satisfying that our state capitol is now the adopted home of the peregrine falcon, one of the world's most spectacular birds. And it is symbolically appropriate that they are nesting below the capitol dome's iconic thunderbird design, symbolizing the Native Americans' belief that the thunderbird will bring life-giving rain to our beautiful land. Our peregrines may not be bringing us abundant rain, but their presence is giving us enormous pleasure.

14

The Ancient Romance of the Yucca and the Yucca Moth

Each June Nebraska's Sandhills and western plains are illuminated for a few short weeks by the stately flowering spikes of Great Plains yuccas, whose spires of twenty to sixty ivory-white flowers emerge from a radiating array of needle-sharp leaves and rise above the rest of the vegetation like Roman candles freeze-framed in flight. Sometimes whole hillsides are transformed by this sudden flower display, which is often variegated with sprinklings of blue spider-worts, golden-yellow hoary puccoons, and white daisy fleabanes. Then, after a few weeks, the spectacular show is over, but for the yucca the most interesting part of the story has barely begun.

Probably most residents of Nebraska can easily recognize a yucca; it is part of a group of about thirty mostly western plants in the lily family having succulent, bayonet-like leaves and huge spikes of large whitish or ivory-colored flowers. The largest of the American yuccas occur in our hottest southwestern deserts. There they may grow to as tall as twenty-five to thirty feet and develop branched, tree-like configurations. Large or small, all yuccas produce spikes of large cup-shaped flowers that open during the night and may remain conspicuous for several weeks.

Yuccas are part of a fairly ancient branch of the plant family tree, perhaps dating back more than fifty million years, when seed plants were first evolving. They have a relatively simple and primitive flower structure, with six conspicuous petal-like structures (three petals, three sepals) of the same color, size, and shape. There is a central, three-parted ovary with a protruding spindle-shaped pistil that extends out from the center of the blossom and ends in a small, indented tip adapted to receiving pollen. Six white stamens surround the ovary, producing packets of sticky pollen at their tips. The stamens are much shorter than the pistil and curve outwardly from it, reducing chances for accidental self-fertilization. The nocturnal flowering, fragrant flowers, and conspicuous white blossoms of yuccas all provide strong clues to the identity of every yucca's pollinator, night-flying moths.

Yucca moths likewise represent one of the oldest and most primitive subfamilies of moths. There are fewer than one hundred known species, and some of the earliest were probably the first moths to develop a nocturnal activity pattern. Most existing species are highly specific as to their food habits, and the single species that pollinates our native Great Plains yucca is especially adapted to feeding upon and reproducing inside this species of yucca.

Some close relatives of yucca moths eat the vegetative parts of these plants, thus injuring them. Some of these very closely resemble true yucca moths and are called "bogus yucca moths." It is possible that ancestral yucca moths operated in this same exploitive way, their larvae feeding on developing yucca seeds, which are highly nutritious and rich in proteins. But, by also eventually assuming pollinating responsibilities for the yucca, the moths assured themselves that there would be abundant seeds available for their larvae to consume. A biologically interesting question obviously follows: How and why did the moth modify its egg-laying behavior in such a way as to also assure the yucca's pollination?

Likewise, how did the yucca transform from a generalized reproductive strategy that probably was originally dependent on wind pollination to one highly adapted for the attraction of and cross-fertilization by a highly specific insect pollinator? By adopting such a risky reproductive mode, the yucca has gained the benefits of precise and efficient cross-pollination, but it also risks losing its entire seed crop to an efficient seed predator. The yucca thereby has not only tied its future survival to that of the moth but also has limited its potential geographic range to that of its single insect pollinator.

As the moth developed new mouth structures and abilities to extract pollen from the stamens, the yucca evidently also modified its pollen characteristics to be more readily gathered by the moth. Equally importantly, the yucca has developed protective adaptations that increase the probability of efficient cross-pollination but reduce the chances of a moth laying so many eggs per flower that all the developing seeds will be consumed by hungry moth larvae.

It is still unknown as to why each moth lays so few eggs per blossom. Typically, each fertilized yucca seedpod chamber holds only one to three moth larvae, leaving the majority of seeds in the pod to mature and eventually be disseminated. It has been suggested that, although each yucca produces several dozen flowers, many blossoms are prematurely dropped, including some that might have already been pollinated. Such apparently random blossom-shedding might force the moth to lay only a few eggs in each of many flowers if it is to increase its chances that at least some of the flowers it pollinates will persist and develop seeds. Not all marriages are made in heaven!

With all this complicated biological background information, it is finally possible to understand the complex story of the moth and the yucca much better. Yucca moths of the Great Plains are so tiny—up to about a half inch long—that they nearly impossible to find. After many long hours of fruitlessly watching yuccas near sunset in hopes of seeing a yucca moth flying in, I finally discovered that by vigorously shaking a flowering stalk I could often cause a moth or two to drop out. In fact, yucca moths often spend the entire day hiding in the half-open blossoms, where their white color renders them almost impossible to see. However, similarly white-colored crab spiders also often hide in these blossoms, waiting patiently for insect prey, which makes yucca blossoms not the safest place imaginable for moths to spend the day!

Male yucca moths are smaller than females, and they lack both an ovipositor and specialized mouthparts that are adapted for gathering and transporting pollen. The moths mate within yucca blossoms as they open, but thereafter the male plays no part in the process. Female moths visit and pollinate flowers between dusk and midnight. She inserts her knifelike ovipositor into the side of the ovary and typically lays one egg in each of the ovary's three main subdivisions or chambers. Before flying to another flower, she scoops up some of the stamens' sticky pollen with her specially modified appendages. She then forms the sticky pollen mass into a ball, which she carries to each of the flowers that she later visits. There she pushes some pollen into the central opening of the flower's stigma and inserts additional eggs, often repollinating the flower between egg-laying bouts. It is not yet known whether a female moth can recognize which flowers have not yet already been pollinated or how many total eggs she might lay at one time. It is known that individual females remain near their "home plant" and are active for less than a week. During this time they presumably survive by eating yucca pollen.

After fertilization the yucca's flowering stalks elongate and enlarge, while large and elongated three-chambered seedpods are formed, each of which is subdivided into two smaller units. During this period the moth eggs hatch and the larvae begin to consume the developing seeds. After it matures over the next month or two, each larva chews its way out of the seed capsule, falling to the ground and pupating in the soil. The adult moths emerge the following spring as ground temperatures warm and the yucca-blooming season arrives.

If too many moth larvae are developing in any single ovary the yucca may abort its seed production. However, in most cases many seeds survive and mature, even in seedpod chambers occupied by two or three larvae, the probable result of egg laying by multiple females. Mature yucca seeds are rounded, flattened, and tightly stacked atop one another, so that several hundred may be present in a single pod. The winged seeds are released when the dried pod splits open and may be carried by wind some distance from the adult plant. Probably only a very few seeds are lucky enough to germinate and survive in the arid environment of western Nebraska, just as very few moth larvae are likely to survive long enough to complete their own life cycle.

This highly unlikely scenario, in which each of two very diverse species depends entirely on the other to achieve its own survival and reproductive success, is sometimes called obligatory mutualism by ecologists. Less technically oriented observers might simply call it amazing.

Tundra swan

15

A Dazzle of Hummingbirds

Query a typical Nebraskan as to whether he or she has ever seen a hummingbird or a UFO, and the response is more likely to be in the realm of flying saucers than flying birds. Ruby-throated hummingbirds are common migrants in eastern Nebraska, and a few stay on to breed along the Missouri River Valley, but it takes special efforts to be able to see them. During May they migrate through the state rather rapidly and rarely stop for more than a day or so at bird feeders before continuing north to begin nesting in the Dakotas, Minnesota, or southern Canada. However, the fall migration is a more leisurely one, with the first adult males usually arriving in mid-August and some females and immature tarrying into middle or late September. It is then that one's best hopes for watching them can be realized.

To achieve this goal, one or more hummingbird feeders are needed, stocked with fresh sugar water. The sugar water should be made at a ratio of one part sugar to four parts of water; it need not be tinted red, although the birds are certainly attracted to red objects. Hummingbirds can detect even small differences in the sugar content of water, at concentrations of up to 20 percent or more, and if given a choice will select the richest source available. As important as putting out hummingbird feeders in a conspicuous location is also having an array of flowers to serve as general attractants. Late-blooming plants with red to orange flowers that are rich in nectar are best for autumn migrants. Flowers with deep, tubular corollas well above ground level that can be reached by hovering and are not surrounded by other vegetation are especially favored.

Trumpet vines are usually at their peak of flowering in mid-August and are great attractants, as are red-colored salvias, but the equally attractive monardas and cardinal flowers evolved to bloom during the hummingbirds' breeding season and are usually past blooming by the time the birds pass through Nebraska. Hummingbirds are known to be able to see into the ultraviolet range of light, so strongly violet-tinted flowers are also attractive to them, but almost any red-colored flower will attract their attention.

August in Nebraska is usually too hot and sultry to venture far outside in search of natural attractions. So it was a welcome surprise for me one day in mid-August a few years ago when three grams of feathered serendipity in the form of a gorgeous male ruby-throated hummingbird arrived in a friend's yard, where there were enough blooming trumpet vines to satisfy any hummingbird's appetite.

We soon put up two hummingbird feeders near a kitchen window, and it took less than five minutes before the hummer had switched from a diet of trumpet vine nectar to one of super-rich sugar water. He quickly made himself at home, perching directly over one feeder and watching carefully for any other intruders. Meantime, I supplemented the feeder with sprays of trumpet vine, to improve its appearance and make any photos appear more natural.

I have tried to photograph hummingbirds for more years than I care to remember, usually being rewarded with no more than shots of bare sky or at most blurred and out-of-focus photos of unidentified flying objects. Then came digital photography, with autofocus telephoto lenses and capabilities of rapid-fire bursts of images. Equipped with a digital camera and telephoto lens, all one now needs is the patience to sit all day, with the camera prefocused, at a hummingbird feeder. One also has to be ready to take advantage of the few seconds it might take for the bird to drink his fill. These visits usually occur about every fifteen to twenty minutes, meaning that over an eight-hour day one might have twenty-five to thirty visits, totaling perhaps only a minute or so of actual photographic opportunities per day.

Hummingbirds are often highly tolerant of people, but this male was skittish of any sudden moves and of shutter noise. He also seemed distinctly annoyed when I covered all but one of the feeder's four feeding openings with tape. This forced him to use the one that was best oriented for me relative to the sun, so that his iridescent gorget might best catch the light in my photos. I also moved one of the two feeders to an inconspicuous spot behind some shrubs, where I thought it would be out of sight, forcing the male to use only the feeder nearest the window. However, it didn't take the male long to find the hidden one, forcing me to hide it even better.

Whenever another male appeared in the yard, the new resident would immediately challenge and engage him in a spirited aerial chase far too swift to follow with the eye. It was never possible to determine the victor with any certainty, but after a male repeatedly returned to a specific perch I assumed that the same bird had always managed to defend its feeding territory. Hummingbirds form no pair bonds, and many species probably distinguish adult males from females simply by the presence or absence of an iridescent gorget. The rules of hummingbird social behavior during the breeding season are seemingly simple—all other males are to be challenged and evicted from the territory, and all females are to be chased and mated.

The intensely brilliant array of highly specialized feathers found in hummingbirds produces some of the purest iridescence to be found in nature, and the gorget is used by males both for courtship and as a threat display toward other males. Males obviously are aware of the visual impact of their gorgets. A common male courtship display is to fly within a few inches of the female, hover in front of her, and, with his iridescent gorget fully spread, quickly shuttle from side to side like a feathered shuttlecock, as if trying to hypnotize her. If mating follows and is successful, the male quickly loses further interest and is likely to turn his attention to other females. Another display technique of many hummingbirds is to perform spectacular power dives from up to about one hundred feet above ground. As the male finally pulls out of the dive, he produces a sudden, sharp noise through feather vibrations or by vocalizing.

Hummingbirds have extremely fine place memory and doubtless can locate favorite feeding locations from year to year, even after a migration of perhaps a thousand miles to wintering areas in Central America and back. Five days after the resident male had arrived at the backyard feeder, a bold female suddenly appeared. It took her only an hour or two to evict the newly settled male, after which she maintained vigilant control over both feeders. In an outright bill-to-bill fight female ruby-throats probably have an advantage over males, as they are slightly larger and about 10 percent heavier on average, but the lighter male is probably swifter and more agile during aerial encounters.

The new female seemed more inquisitive than the male and would sometimes fly up to the window where I was sitting, trying to get a good look inside. She was more prone to try to get nectar from the trumpet-vine blossoms I put out, although both sexes were more interested in probing freshly picked blossoms that angled slightly upward than in older, more dangling blossoms that had wilted somewhat and whose nectar had probably drained out. The female also was less bothered by my camera movements and shutter noises than the male had been. As a result, I was finally able to get some of the photos that I had imagined for many years.

In eastern Colorado and Wyoming bird-watchers and photographers are unlikely to see ruby-throats but have greatly increased chances of seeing broad-tailed, rufous, and calliope hummingbirds. The broad-tailed is similar to but slightly larger than the ruby-throated, and males produce a distinctive loud buzzing whistle when in flight. Broad-tails are common from May to September along the Front Range foothills and lower mountains north to southern Wyoming. From July to September fall-migrant rufous hummingbirds are also common at the same elevations, with males arriving first, followed by females and finally by juveniles. All ages and both sexes of this species are strongly rufous on the upperparts and underparts, as well over most of the tail.

The rufous hummingbird breeds as far north as Anchorage, Alaska, and often winters in western Mexico, well over one thousand miles away, and sometimes strays to the Gulf Coast or even the Atlantic coast, some two thousand miles away. Considering that a rufous hummingbird might live as long as ten years and migrate at least two thousand miles per year, it is possible that some hummers fly a distance equivalent to traveling around the world at the equator in the course of a lifetime.

Calliope hummingbirds are the smallest of the North American hummingbirds, averaging only a tenth of an ounce, and are common in western and northern Wyoming mountain ranges but rare in Colorado. Although only about three inches in length, the male calliope has a stunning iridescent red gorget of elongated feathers. Campers in the Rocky Mountains often carry hummingbird feeders with red spouts to attract hummers. I have a green tent with ornamental red striping; this pattern alone is sometimes enough to attract them to the tent.

I will never forget my many observations of calliope hummingbirds during two summers that I spent in Grand Teton National Park. When I regularly entered a male's territory near camp, he would apparently treat me like a rival, perhaps because I often wore a red cap. He would fly to a height of about thirty feet, then hover overhead and orient himself relative to the sun so that his expanded ruby-red throat gorget was pointed directly at me. The resulting visual effect was that of a laser-like beam of light coming down out of the sky—an impact that certainly would not be overlooked by either visiting females or intruding males.

Western hummingbird species sometimes track well to the east of their usual migration routes, often passing through eastern Nebraska, and have even appeared as far east as the New England states. Short of traveling to Latin America, where most of the family's three-hundred-plus species are found, the best region for seeing North American hummingbirds is in the Southwest, especially Arizona, where at least six species might possibly be seen in a single day.

Watching hummingbirds is almost like being given the opportunity to observe a totally different world from our own, where time is perceived in milliseconds, and observing them may force us to think of the amazing amount of intelligence, memory, and geographic information that is packed into a brain far smaller than a pea. A botanist might ponder the relative importance of nectar production, flower color, flower shape, and blooming times in shaping the evolution of hummingbird-pollinated flowers, while an ornithologist might wish to study the anatomical and foraging adaptations that make hummingbirds perfect pollinators, favoring the long-term survival of both bird and plant. Or one may simply relax, watch hummingbirds, and revel in their dazzling beauty.

16

A Symphony of Swans

Because of their immaculate white plumage and their strong pair and family bonds, swans have also long served as icons of beauty, devotion, and longevity in the myths and folklore of many cultures. Our personal interests in and perceptions of wild swans are often formed in childhood, by reading such classics as Hans Christian Anderson's stories "The Ugly Duckling" and "The Wild Swans" or E. B. White's *The Trumpet of the Swan* or perhaps upon seeing a performance of Tchaikovsky's *Swan Lake* ballet.

Many of the most admired human traits, such as permanent pair bonding, extended biparental care, and family cohesion, are biological facts in swans, but the sadly romantic idea of a dying swan uttering a final "swan song" is only folklore. Yet a famous American biologist, D. G. Elliott, reported in 1898 that once, after he had shot and wounded a whistling swan in flight, it began a long glide while uttering a series of "plaintive and musical" notes that "sounded at times like the soft running of the notes of an octave" as it gradually drifted downward. Nowadays such unusual behavior would probably be interpreted as only an instinctive distress call, but Elliott's story might have provided an early factual basis for this commonly used expression.

Most Americans are probably personally familiar with the regal-looking mute swan of Europe, which has long been imported to American parks and zoos. Mute swans have also long been used by wealthy landowners to decorate private ponds and help control the growth of unwanted aquatic plants. When mute swans escaped from Long Island estates during hurricanes in the late 1930s, many became feral, and their offspring have since expanded over much of the Atlantic coast, from New Hampshire south to the Carolinas. Following introductions dating back to 1919, mute swans have also spread out from the eastern shoreline of Lake Michigan, occupying much of the Great Lakes region from Wisconsin to New York, and are now considered an invasive species in several states. A few seemingly wild mute swans have been seen recently in Nebraska, but the evidence as to their origins is still too murky to add them to the official list of Nebraska birds.

Two other swans of comparable beauty but unquestionably wild origins can be seen in Nebraska. Unlike the virtually silent mute swan, both of our native American swans have loud, clarion voices that were the historic basis for their English names, whistling swan and trumpeter swan. The smaller (up to twenty pounds) whistling swan, now officially called the tundra swan, has a musical, more soprano-like, voice, while the trumpeter's is louder and more baritone-like. During the 1980s the English name tundra swan officially replaced whistling swan, to describe its high-arctic breeding habitat and to include within the same species its close Eurasian relative, the Bewick's swan.

The trumpeter swan is a substantially larger species, with adults sometimes reaching thirty pounds, making it the heaviest of American birds. It too has a very close and slightly smaller Eurasian relative, the whooper swan, but so far these have been considered biologically distinct species. All four of these swans have loud, clarion voices and often use them in long-distance communication and territorial interactions.

These vociferous swans are also notable among waterfowl in having windpipes (tracheae) that penetrate the bony breastbone and form a long internal loop within it. Just before the windpipe enters the lungs, it is transformed into a bony sound box (the syrinx) with paired vibratory membranes that are set into motion when the lungs expel air. Varied tensions on the paired membranes influence the rate of their vibrations. This vibration rate sets the basic frequency of the resulting sounds that, because of subvibrations, produce additional overtones (harmonics) that enrich the vocalizations. The windpipe further modulates and amplifies particular sound frequencies, depending largely on its volume and length, enhancing harmonic complexity. These remarkable adaptations in trumpeter and tundra swans allow for a great range of individual vocal variations and a high degree of harmonic development, making every individual swan's voice unique and probably easily recognizable by others. Very similar structural adaptations are present in whooping and sandhill cranes and produce similarly loud and individually identifiable vocalizations.

As high-arctic nesters, tundra swans appear in Nebraska only during spring and fall migrations. During fall migration, those swans breeding in the central Canadian Arctic fly south to North Dakota, gradually veering toward the east as they approach South Dakota. The swans then follow a southeastern route roughly paralleling the Minnesota River Valley and from there continue eastward to wintering areas that extend from Chesapeake Bay south to the Carolinas.

This diagonal migration route means that only in northeastern Nebraska is one likely to encounter tundra swans, while they pass southward during November–December and return in March–April. However, in North Dakota and eastern South Dakota tundra swans are sufficiently common that so-called sport hunting is permitted. At least four thousand tundra swans of the eastern North American population are legally killed annually in the Dakotas, North Carolina, and Virginia. Several thousand more are killed by poachers or are wounded but never retrieved, with the total kill probably approaching ten thousand birds annually, or perhaps 10 percent of the eastern population.

Swans from the western population that breeds in western Canada and Alaska take a different fall migration route, which passes from Saskatchewan and eastern Alberta southwest through Montana, Utah, and Nevada and then on to California wintering grounds. Like the eastern flock, the western North American population probably numbers at least one hundred thousand birds, and they can be legally hunted in states such as Montana, Utah, and Nevada. Hunting in those states, plus subsistence hunting by Natives in Alaska, probably results in the deaths of at least ten thousand swans annually, also about 10 percent of the total western population.

Luckily, America's largest swan, the trumpeter, is fully protected. It was on the list of federally endangered species for many years and was not removed from that list until it became apparent that a large and previously unstudied population of swans in southern Alaska are actually trumpeter swans rather than tundra swans. Additionally, since the 1960s great efforts have been made to relocate trumpeter swans from surviving populations in the Rocky Mountain region to historical breeding areas of the northern Great Plains, substantially increasing the species' total population, which by 2012 numbered over forty-six thousand birds.

Nebraska's historic breeding trumpeter swans were extirpated by 1900 but were later reestablished during the 1960s as a result of releases of cygnets at Lacreek National Wildlife Refuge, South Dakota. Restoration efforts there, at the northern edge of the Sandhills, were so successful that by 1987 the population had reached nearly 300 individuals. Expansion southward into the Nebraska Sandhills followed, so that by 1999 there were nearly 600 records for the Nebraska Sandhills, with Cherry and Garden Counties accounting for about 70 percent of the total, as well as 160 nesting records. Nesting now occurs in many Sandhills lakes and large, shallow marshes. Additionally, some spring-fed streams of the Sandhills, such as the North Loup River and Blue Creek, usually remain unfrozen all winter, eliminating the need for long seasonal migrations. The current Nebraska breeding population probably numbers several hundred.

Trumpeter swans mate permanently, and each pair returns to its nesting area in spring as soon as the weather allows. Nesting marshes in Nebraska are typically large, shallow, and well vegetated, with abundant shoreline plants and submerged aquatic vegetation. Marshes having muskrat present are favored, as their "houses" provide a convenient nest substrate, protected from most wave action. Nesting territories average more than thirty acres and sometimes exceed one hundred acres. They are vigorously defended, the adults even excluding their own offspring from previous years. The male performs most territorial defense, but after territorial disputes the female participates in "triumph ceremonies" that are marked by loud mutual calling and wing waving. She also helps defend the nest site when needed.

Both sexes help construct the often-bulky nest, which may simply be the flattened top of a muskrat house or may consist of piled-up reeds and bent-down emergent vegetation that provide an elevated platform. Nesting behavior in the Sandhills has been seen as early as April 28. The eggs (typically four to six) are then laid at two-day intervals, with incubation starting only after the clutch is complete. The female performs most incubation over the thirty-two-to-thirty-seven-day period to hatching, while the male patrols the territory.

Hatching in Nebraska is likely to occur during late May or early June. The cygnets hatch within a few hours of each other and are led from the nest within twenty-four hours of hatching. The cygnets' fledging period to initial flight is approximately one hundred days, which means that the brood's first flights might not occur until September. The cygnets remain with their parents for at least their first year of life but are evicted from the nesting territory by the following spring. From four to seven years might pass before the young birds begin to breed, although sexual maturity is reached much sooner.

As a protected species, trumpeter swans often live for ten years or more and are known to have survived for at least twenty-four years. Wild tundra swans appear to have somewhat shorter life expectancies that rarely exceed ten years, perhaps in part because of hunting-related mortality, their much longer and more stressful migration routes, and the rigors of arctic nesting.

Some of the places along Nebraska highways where trumpeter swans can usually be seen during summer are on larger marshes off U.S. Highway 2 in Grant and Sheridan Counties, at a wetland off the South Loup River at the southern edge of Ravenna, and near U.S. Highway 83 in Cherry County, between Valentine and Valentine National Wildlife Refuge. In 2011 four pairs produced eleven cygnets at Valentine National Wildlife Refuge, making it one of the state's premier swan-breeding sites.

Continued releases of trumpeter swans in the Midwest, especially in Minnesota, have greatly increased the chances of seeing trumpeter swans in the region. By 2012 there were perhaps ten thousand trumpeter swans in the eastern North American population, breeding from South Dakota and Iowa east to Ontario. Because of highly successful restoration efforts in Minnesota (with a 2012 population of fifty-five hundred birds), thousands of trumpeter swans migrate through Iowa each spring and fall, supplementing that state's own recently established population of breeding birds. Some of the places in Iowa where migrating trumpeter and tundra swans can usually be seen include Forney Lake Wildlife Area near Thurman, Union Slough National Wildlife Refuge near Bancroft, and the Upper Mississippi National Wildlife and Fish Refuge near McGregor.

One of the increasingly important regional wintering areas for trumpeter swans near Nebraska is Squaw Creek National Wildlife Refuge, near Mound City, Missouri. From November through winter as many as 456 swans have been seen, as well as up to a half million or more migrating snow geese and countless other waterfowl. It provides a visual spectacle that one is likely to remember and cherish for a lifetime.

17

A Plethora of Pelicans

Of all the birds of the Great Plains, probably none is more widely recognized by the general public than the American white pelican. Its almost cartoonlike profile, with an impossibly long, pouched bill and a waddling gait on land that reminds one of an overweight uncle, is impossible to mistake. Yet, when swimming, a group of pelicans has the appearance of a slow but regal procession, with each bird maintaining a decorous distance behind the leader. In flight the birds produce a mesmerizing slow-motion aerial ballet. With their great wing areas providing lift for their relatively heavy bodies, pelicans can soar effortlessly in the slightest updraft, and during normal flight nearly a third of their time is spent in restful gliding. To maximize flying efficiency, the birds often assume an echelon formation, with the lead bird setting the flap-glide rhythm. As the lead bird shifts from flapping to gliding, the bird just behind does the same, and the rest follow in rapid progression. The result is a wavelike, almost hypnotizing visual effect, which reduces the energy cost of flying for all of the birds using the slipstream of the one just ahead.

White pelicans have long been a prominent part of the Great Plains scene. When traveling up the Missouri River in August 1804, along what is now the Iowa-Nebraska border, Lewis and Clark came upon a flock of perhaps as many as five thousand to six thousand pelicans. One was shot, and, as if to prove Dixon Merrit's familiar limerick about pelican beak capacity, its pouch was found to hold up to about a gallon of water!

Pelicans use their huge beak pouches to capture fish with an open-mouth stabbing movement, taking in both water and prey. They then shut the beak, forcing water out its sides and trapping any animals inside. Pelicans often feed alone by this method, but as with some other fish-eating birds such as cormorants, coordinated group fishing is a more efficient way of foraging. Working in groups of two to six birds, pelicans will typically use a semicircular formation to drive any fish into fairly shallow water, and, as if on command, all will suddenly thrust their beaks into the water in unison. The groups that are the most coordinated in this regard are the most efficient in catching fish.

Most pelican prey consists of fairly slow-moving fish, such as bullheads, carp, and suckers, but salamanders have been found to be an important food item in North Dakota and crayfish are sometimes also captured. In spite of their huge pouches, the tongues of pelicans are extremely small, and they seem to eat almost anything they can capture and swallow.

In spite of the birds' catholic diet, fishermen have long despised pelicans, and up until the mid-twentieth century their continental population was in decline. Even in Yellowstone National Park, the unofficial (and illegal) policy was to raid the Yellowstone region's only nesting pelican colony and destroy their eggs, in a senseless effort to try to protect the park's cutthroat trout population for exclusive use by fishermen. Luckily, that practice has changed, and since the 1960s the pelican's continental population has been increasing at an estimated average annual rate of 3 percent.

Although the closely related brown pelican is a coastal species, the American white pelican breeds only in the continental interior and reaches its highest breeding abundance in the northern Great Plains, especially in the shallow lakes and "pothole" marshes of Canada's prairie provinces. No recent nationwide population surveys are available, but in the 1980s the United States supported an estimated twenty-two thousand nesting pairs and Canada over fifty thousand pairs. Given the recent population trend, by now these numbers have probably doubled.

Persons wanting to see pelicans in Nebraska have many opportunities. Pelicans are fairly regular migrants on the wider rivers, just as they were in Lewis and Clark's time. The central Platte Valley, where the river channel is widest and where there are few shoreline trees to obscure the view, is especially attractive. In a 1990 study, John Sidle and others found that migrating pelicans occurred there in the largest numbers where the river channel was at least eight hundred feet wide.

Lakes and reservoirs, especially their shallower portions, are equally attractive to pelicans. During May I have seen as many as seven hundred to eight hundred along the western end of Calamus Reservoir, and comparable numbers occur near the western end of Harlan County Reservoir. Pelicans are also common at Crescent Lake and DeSoto National Wildlife Refuges during migration. I have seen pelicans on Lake McConaughy on almost every visit from spring to fall. Some nonbreeders remain there through the summer breeding season, often "hanging out" below Kingsley Dam on Lake Ogallala, waiting for stunned fish to come through the dam's outlet and provide an easy meal.

Some locations near Nebraska that support breeding pelicans include Lacreek National Wildlife Refuge, at the northern edge of the Nebraska Sandhills near Martin, South Dakota, and other parts of northeastern South Dakota (in Roberts, Marshall, Day, and Coddington Counties), which may have small colonies. Pelicans also breed locally or periodically in southeastern Wyoming and regularly in southwestern Minnesota, where Marsh Lake (Lac qui Parle County) had 1,450 nests in 1983. North Dakota has a single but very large breeding colony at Chase Lake, in Stutsman County. Breeding records there go back at least to 1905, and as many as 4,100 nests were present during the early 1960s.

Pelicans become sexually mature as three-year-olds, so many yearling birds and some two-year-olds often spend their summers at points between their Gulf Coast wintering grounds and their Great Plains breeding grounds, from South Dakota to central Canada. Most of these birds lack the horny enlargement present on the top of the upper bill that is typical of breeding birds. This curious structure, resembling an inverted keel, is chemically like the keratin in human fingernails and varies considerably in shape. It develops every spring on adults of both sexes and drops off near the end of the nesting period. Like driving a red sports car, it is evidently a visual signal of breeding readiness, but it seems to have no other apparent function. Visual social signals might be especially important to pelicans, as they are nearly mute, and in several species the beak and pouch coloration becomes much more colorful during the breeding season.

Spring migration begins shortly after rivers and lakes thaw and is aided by the warming spring atmosphere and the associated thermals, which provide strong updrafts. Pelicans are slow fliers, averaging about thirty miles per hour, and may not arrive at their northern breeding grounds until early May. They typically fly in flocks of up to about 250 birds, possibly covering up to a hundred miles or more in a day. Once on the breeding grounds, adults may make daily trips to favored feeding areas that may be as far as forty miles away, or rarely even as much as sixty miles.

Shortly after arrival on the breeding area the birds begin courting and looking for nesting spots. Pelicans typically nest on low, flat islands that can easily be flooded during storms. They are not known to maintain their pair bonds over winter, but monogamous pair bonds are quickly formed and last until the young have fledged. Two eggs are laid, and thereafter both sexes take turns in incubation, which lasts about thirty days. Since incubation starts with the laying of the first egg, the second-laid egg produces a chick that is hatched a day or two after the first.

The young are hatched naked and blind, and initially they more closely resemble reptiles than birds. The first-hatched chick grows rapidly, being provided with a nearly continuous supply of half-digested fish from both parents. Its sibling, however, is less fortunate, and as a result of parental neglect and aggression on the part of the older chick is likely to starve to death in less than two weeks. Typically fewer than 10 percent of successful broods still contain two chicks at the time of fledging, which occurs at ten to eleven weeks of age.

The fall migration is a leisurely one. Migrant birds typically arrive in Nebraska by late September, and historically they were gone from the state by early November. The warming climate trend of recent years may soon lead to later fall migrations and earlier spring appearances.

Since 2011 a celebration of the pelicans' spring return to southern Nebraska has developed at Harlan County Reservoir, with a White Pelican Homecoming Celebration held at Alma during late March. It is typically held during the last week of March, with a Harlan County White Pelican Watch extending from March 1 to mid-April.

There is something uniquely relaxing about watching a flock of pelicans in flight. At times they might circle slowly, safely within the grasp of an invisible thermal, perhaps gaining a thousand feet or more of altitude in the manner of cranes, until they almost disappear from sight. At other times pelicans resemble nothing so much as a regatta of sailboats on royal parade, silently riding on the wind in measured procession. Unlike the cranes with their excited clamoring, or the snow geese with their incessant dog-like yelping, the pelicans simply fly silently on, as if they were aware that to add sound to their presence would in some way only lessen its impact.

WINTER

Golden eagle

18

A Gathering of Eagles

Nebraska is blessed in having substantial populations of both species of North American eagles, the bald eagle and golden eagle. The bald eagle, our national symbol, has become sufficiently common during the last four decades that it is not unheard of to see them perching or fishing within the city limits of Lincoln. In the summers of 2012 and 2013 a pair even nested along Salt Creek, at the northern edge of the city. Yet from 1962, when I first arrived in Lincoln, until the late 1970s a sighting of bald eagles almost anywhere in the state would be memorable.

Only after the use of DDT and other chlorinated hydrocarbons to control insect pests was outlawed nationally did the bald eagle begin to recover from a population slide. The decline had resulted from DDT's physiological effects on the bird's reproduction system, upsetting its calcium metabolism and causing it to lay weakened, almost calcium-free eggs. The female's resulting eggshells were so thin that they were often crushed by the weight of the incubating bird or the embryo was unable to extract enough calcium from the shell to develop bones strong enough for the chick to survive. The presence of DDT had devastating effects on reproduction in many species of birds, especially fish-eating species such as ospreys and other predators at the top of food chains, such as peregrine falcons.

By the early 1970s the only population of bald eagles in the United States apparently unaffected by DDT was Alaska's, which was still relatively pesticide-free. In 1978 all other populations of bald eagles in the coterminous United States were placed on the list of nationally threatened (Michigan, Minnesota, Oregon, Washington, and Wisconsin) or endangered (all other states) species. By the late 1970s the U.S. population appeared to be on the increase, and as of the early 1980s there were an estimated eighty thousand birds present in North America, with about forty-eight thousand in Alaska and Canada and about twenty-two thousand everywhere else.

The bald eagle was recently removed from the federal threatened and endangered list and has since been thriving. Considering continent-wide estimates derived from the North American Breeding Bird Survey, the average annual rate of significant population increase of the species between 1966 and 2011 was 4.9 percent, placing it among the ten most rapidly increasing bird populations of North America.

Nebraska's breeding and wintering bald eagle populations have correspondingly increased, but few statewide surveys of winter populations have been attempted. Statewide midwinter counts made by the Nebraska Game and Parks Commission and the National Wildlife Federation between 1980 and 1998 averaged 743 birds, with a maximum of 1,340 seen in 1999.

Although bald eagles concentrate on many of Nebraska's reservoirs during winter, the largest numbers are to be found around Kingsley Dam, about eight miles north of Ogallala, and its 35,700-acre-capacity Lake McConaughy reservoir and small (640-acre) outflow pool (Lake Ogallala/Keystone). There is now an eagle observation building overlooking Lake Ogallala, constructed by the Central Nebraska Public Power and Irrigation District. It is open from January 7 through late February from 8:00 a.m. to 4:00 p.m. MST. The largest eagle numbers are typically present in January and February, but daily numbers vary greatly, and up to nearly four hundred have been recorded in a single day. Information on eagle numbers, and related birding information, can be obtained from the Kingsley Dam office at (308) 284–2332.

The Missouri River Valley and other larger state reservoirs, such as Calamus and Harlan County, also attract substantial numbers of migrating or wintering eagles, as well as occasional nesting birds. During the three most recent years of Christmas Bird Counts, the Lake McConaughy count total has ranged from 52 to 73 bald eagles, as compared with overall state totals that varied from 209 to 267 individuals. Missouri River corridor counts, represented by those from Omaha and the DeSoto and Boyer Chute National Wildlife Refuges, have had numbers mostly only somewhat lower than recent counts from Lake McConaughy.

Bald eagles had vanished as a breeding species in Nebraska and over much of the Great Plains by about 1900, and it was not until 1973 that the first instance of attempted but unsuccessful in-state nesting was recorded in Cedar County. A few other records of unsuccessful nesting efforts were obtained during the 1980s in Garden, Lincoln, and Saunders Counties, but it was not until 1991 that the first known successful nesting occurred, in Douglas County. After that the number of active nests in Nebraska rose to ten by 1996, twenty by 2000, and fifty-five by 2007.

Since then nesting numbers have fluctuated but have been recorded almost statewide. By 2009 nests had been noted in sixty-two of Nebraska's ninety-three counties, according to a 2010 report by Joel Jorgensen and others. Most nest have been reported along riparian corridors in the northern and eastern parts of the state. Between 1973 and 2009, 85 percent of 221 nests were situated along rivers, especially the Platte (42), the Missouri (37), and the Elkhorn (15). Of 440 active nests observed between 1991 and 2009, an average of 1.48 eaglets fledged per nest, with nest losses caused by strong winds being a significant factor in breeding failures. This nest-success incidence is considerably higher than the rate of 1.03 young fledged per active nest reported from four increasing populations in the greater Yellowstone region of Wyoming.

The situation of the golden eagle in Nebraska is quite different. Unlike bald eagles, golden eagles are much more common in the water-scarce and relatively treeless parts of the state, and especially in the bluff and canyon country of the northwestern Panhandle. There, their most favored prey, prairie dogs and jackrabbits, are most likely to be found. Nesting records prior to 1960 are mostly confined to Sioux, Scotts Bluff, Dawes, and Garden Counties, plus two isolated records for Keith and Douglas Counties. Post-1960 records exist for nearly every Panhandle county, as well as for Keith, Lincoln, Chase, and Hayes Counties. During the first Nebraska Breeding Bird Atlas studies of the middle to late 1980s, all of the possible to confirmed nests were confined to the Panhandle, and the only two confirmed nestings were in Banner and Cheyenne Counties.

Nationwide, golden eagle populations appear to be stable or slowly declining. Breeding bird surveys from 1966 to 2009 indicated a statistically insignificant population trend. However, in the Intermountain West where golden eagles are most abundant, clearing of sagebrush and conversion of native grasslands to irrigated farmlands have probably caused widespread population losses. The continuing uncontrolled poisoning and persistent slaughter of prairie dogs, often by government agencies charged with protecting our natural resources, has additionally had serious secondary effects on golden eagles, swift foxes, ferruginous hawks, prairie falcons, black-footed ferrets, and other predators that depend upon them.

In spite of their federal protection, some eagles are still being purposefully killed. Among nearly 5,000 hawks and eagles handled by the Nebraska Raptor Recovery Center between 1976 and 2011, there were 239 bald eagles and 102 golden eagles, many of which had been wounded by trigger-happy hunters. Some eagles also are fatally poisoned by ingesting lead pellets imbedded in a prey animal or by scavenging a poisoned carcass. Seeing the lights gradually fade from the eyes of a poisoned golden eagle as it was dying at a raptor recovery center brought tears to my eyes, and the memory of it still haunts my thoughts every time I see a wild golden eagle. The experience made me remember Aldo Leopold's confession about seeing the fading of the "fierce green fire" in a eyes of a gray wolf that he had just shot while he was working as a Forest Service employee in New Mexico.

Where to See Eagles in Nebraska's Surrounding States

South Dakota. All of the larger Missouri River reservoirs (Lake Francis Case, Lewis and Clark Lake, Lake Sharpe, and Lake Oahe) attract eagles during colder months. Karl Mundt National Wildlife Refuge, located just below Fort Randall Dam, is an important bald eagle wintering site, and there is an eagle-watching platform at the dam. Golden eagles breed widely but locally in western South Dakota, especially in the rimrock-edged canyons of the Black Hills and Badlands. They move into lower grassland areas during winter, where jackrabbits are most common.

Iowa. Bald eagles nest locally but increasingly commonly in Iowa, mainly near larger rivers such as the Missouri and Mississippi, including the Upper Mississippi National Wildlife and Fish Refuge, headquartered at McGregor. There they are very common in late fall, especially just before freeze-up. In western Iowa they migrate along the Missouri River, wintering wherever open water allows, with large numbers visiting DeSoto National Wildlife Refuge near Missouri Valley when goose populations are at their peak. Golden eagles are rare winter visitors in Iowa, most often being seen in northeastern Iowa (Allamakee and Scott Counties) along the Mississippi River, in company with bald eagles.

Missouri. Up to several hundred bald eagles congregate at Squaw Creek National Wildlife Refuge, near Mound City, in late fall (typically peaking in early December), and there is usually a nesting pair or two on the refuge. Elsewhere in the state they now breed regularly and widely, following restoration projects begun in Wayne and Vernon Counties, and they winter commonly along rivers, larger wetlands, and lakes, such as at Swan Lake National Wildlife Refuge, near Sumner, and Mingo National Wildlife Refuge, near Puxico. Golden eagles are very rare in Missouri.

Kansas. Bald eagles nest locally in eastern Kansas, west to Hodgeman County, usually near lakes or reservoirs. During winter large numbers occur on the reservoirs and rivers of northeastern Kansas, such as Tuttle Creek, as well as along the Arkansas River and at Cheyenne Bottoms Wildlife Area and Quivira National Wildlife Refuge, both near Great Bend. Golden eagles are very rare summer residents in the west and are uncommon migrants and winter residents in western grasslands, especially near prairie dog towns.

Colorado. In Colorado bald eagles are common during winter in western valleys, mountain parks, and wetlands. During migration many concentrate at the eastern reservoirs along the South Platte River, such as Jumbo Lake, near Crook, and Prewitt Reservoir, near Sterling. Historically most breeding by bald eagles occurred in northwestern and southwestern Colorado, but it is much more widespread now. Nesting by golden eagles is mostly limited to northwestern Colorado, and this region also attracts the most golden eagles in winter.

Wyoming. There is also a probably expanding nesting population of bald eagles in Wyoming, largely distributed along western rivers but with records extending to the state's eastern borders. Bald eagles often have formed substantial winter concentrations in river canyons such as those of the North Platte, including Jackson Canyon, near Casper, and Boxelder Canyon, near Glenrock. Wyoming perhaps supports the largest breeding golden eagle population of any state south of Alaska; in 1973 the state's population was estimated at ten thousand birds, and breeding records exist for every county. With the increasing destruction of the sagebrush ecosystem in Wyoming, jackrabbits (the golden eagle's primary prey) are likely to be declining, so the long-term future of the golden eagle in Wyoming is problematic.

19

A Parliament of Owls

Few people are entirely neutral as to their attitudes about owls. As mysterious nocturnal visitors, owls' voices send fear into the hearts of many but offer a haunting and relaxing mantra to others. Probably many people go their entire lives without ever seeing a wild owl; their camouflage-like plumages blend into their background so well that a family of owls can exist in a hollow tree in one's backyard without even being noticed.

The winter of 2011–12 provided an exception to this general rule. Snowy owls, the only owls in the world that are almost entirely white and are thus perfectly hidden in their tundra environment, staged a rare invasion into the northern and central states, where they stood out like white flags against a dead-grass background. The reasons for this incursion are still uncertain, but a shortage of their primary arctic food base (mouse-like rodents called lemmings) seems a possible explanation.

The rarity of this invasion is reflected in the number of snowy owls handled at Fontenelle Forest's Raptor Recovery near Bellevue, where sick and otherwise disabled birds of prey are treated. During the thirty-five-year period from 1976 to 2011, a total of only twenty-one snowy owls had been treated by the center. However, between December 2011 and mid-January 2012, ten more snowy owls were brought in. Nine of these were young, highly emaciated birds that were extremely weak and could not be saved. These birds probably lacked the experience to hunt successfully during times of low food supplies and had been forced southward in search of prey.

Given the publicity and interest produced by this nationwide snowy owl invasion, a survey of the other owls of Nebraska might be of interest. Besides the snowy owl, two other Nebraska owls are migratory. The northern saw-whet owl is a small, inconspicuous bird that also mainly breeds in coniferous forests to the north of Nebraska. It is an uncommon permanent resident in the Black Hills of South Dakota and has been suspected of nesting in the Pine Ridge region of Nebraska's Panhandle. Otherwise, it is a regular fall and spring migrant in the state, perhaps locally overwintering. Although the name "saw-whet" refers to the supposed resemblance of its somewhat raspy note to that of a saw being whetted, its vocalizations more often consist of simple repeated whistles. Like our other mostly high pitched nocturnal owls, it has a very large, disk-shaped group of feathers surrounding each ear, providing two parabola-like sound receptors and making it a highly effective auditory hunter under conditions of nearly total darkness.

Our other migratory owl is the burrowing owl, a familiar sight in western areas where prairie dogs still survive. Unlike our other owls, it is active during the daytime, and it can often be seen standing sentinel-like near its nesting cavity, a behavior that has inspired one of its appealing colloquial names, "howdy owl." It often nests in abandoned cavities among prairie dog colonies, although almost any appropriately sized cavity seems to work just as well. When nesting among prairie dogs, it seemingly maintains a tolerance for and an uncertain peace with them, being too small to represent much of a threat. Although burrowing owls often prey on small rodents such as mice, much of their summer foods consists of insects, especially large beetles. By late September, as the insect supply dwindles, the owls begin to depart for warmer wintering areas, probably migrating as far south as western Texas.

Slightly smaller than the burrowing owl, our eastern screech-owl is perhaps the most common owl in the state, at least in wooded habitats such as river valleys, suburban parks, or even well-treed residential neighborhoods. Its call, a soft, tremulous whinny, would probably not be recognized by most people as coming from an owl, and it is small enough to nest in a woodpecker's cavity. About 90 percent of the eastern screech-owls in Nebraska are of the gray plumage type; the remainder are a rich rufous color. The latter plumage variant is progressively more common in the eastern states, where it may be less conspicuous than gray plumage among the brighter colors of the deciduous forests there. In the shady coniferous forests of the far west only gray-plumaged birds are present.

Nebraska's medium-sized owls are the barn owl, barred owl, short-eared owl, and long-eared owl. Two of these, the barn owl and barred owl, are highly nocturnal. Relative nocturnality is somehow correlated in owls with a dark iris coloration. Owl iris color probably has nothing to do with visual sensitivity in owls, but yellow eyes are highly conspicuous during daylight and may serve as effective social signals, whereas brown eyes are not. Owls in general, and certainly all nocturnal owls, have eyes with extremely high sensitivity in low light (rod-based or monochrome vision) but little or no capability for detecting different hues (cone-based color vision). Cone-based vision allows both for fine color discrimination and high-resolution imaging but requires a fairly high light level to be activated.

Of the four species just mentioned, the short-eared owl and barn owl are open-country owls; the long-eared owl and barred owl are forest-dwelling owls. The barn owl is the only member of a group of similar owls that occur around the world, especially in tropical regions, and are noted for their extraordinarily keen hearing abilities.

Barn owls have a remarkable ability to locate and unerringly catch small rodents in total darkness. Like saw-whet owls, they have large facial disks of feathers around their ears, and, like most mammals, their ears have external flaps that can help to capture and funnel extremely weak sounds into the inner ears. Barn owls are also unique in producing a human-like screaming call that, when heard on dark nights, might help one understand the widespread if unjustified fear of owls.

In contrast, the long-eared, short-eared, and barred owls have typical owl-like hooting calls that differ in cadence characteristics and allow for distinction on that basis alone. The barred owl is easiest to recognize; its call cadence can easily be remembered as, *Who cooks for you? Who cooks for you-all?* The short-eared owl utters a series of raspy barking notes, and the long-eared owl produces occasional long, soft, and usually single hoots. All four of these middle-sized owls are effective mouse and rat killers, and any of them would be a very good neighbor for anyone with rodent pest problems.

The great horned owl and snowy owls are the largest of our regularly occurring owls, with great horned owls often weighing more than three pounds and snowy owls at least four pounds. In both species, and in predatory birds generally, females are considerably heavier than males on average, perhaps because females undergo the additional stress of egg laying and their greater body mass improves their chances of survival.

Great horned owls are the most opportunistic predators of all Nebraska's owls. They will not hesitate to take on prey weighing as much or even more than themselves, such as striped skunks, which may weigh up to about ten pounds. Great horned owls have even been known to attack domestic cats and very small dogs.

Although snowy owls are the heaviest of all North American owls, their usual lemming prey weigh only about two to five ounces, or roughly 5 percent of the owl's body weight. Thus many lemmings per day would be needed to avoid starvation, so a lemming shortage might have serious repercussions for snowy owl populations. Migrating and wintering owls prey on any mammals and birds that might be locally available, up to the size of hares and pheasants.

Over the thirty-five-year period from 1976 to 2011, Raptor Recovery Nebraska treated a total of 4,999 variously disabled owls. In descending numbers, they included the great horned owl, eastern screech-owl, barn owl, barred owl, long-eared owl, short-eared owl, burrowing owl, snowy owl, and northern saw-whet owl. Persons finding disabled owls, or any raptors needing help, are urged to contact Fontenelle Forest's Raptor Recovery (402-731–3140) or any local representative of Nebraska Game and Parks. At least twenty-five towns have local volunteers who can receive and transfer injured birds to the center, as do Nebraska Game and Parks personnel. It should be remembered that all raptors have powerful talons and razor-sharp beaks, so they should never be handled by people lacking experience.

One of our least-known groups of native birds, the owls are among the most fascinating. Their hearing abilities are perhaps the finest of any birds in the world, and their nocturnal vision is also unsurpassed. Even their daylight vision is almost unbelievable; the great gray owl is able to detect a mouse from more than a hundred yards away. Any owl is a puzzle enclosing an enigma, both of which are enveloped in the softest and most beautiful plumage imaginable. What bird could be more intriguing than an owl?

20

The Feathers of Winter

For many Nebraska birders, the last big event of the year is the Audubon Christmas Bird Count, which is held annually during the last week of December. It is an occasion to join with friends in a day out to try and see as many species as possible in a single day. More importantly, it produces information that, combined with that from more than fifty thousand other observers, provides a highly documented population sample of early winter birds throughout North America, Latin America, and the Caribbean region. The tradition began in 1900, and as of 2011 there have been 111 national counts. Nebraska counts began in Lincoln and Omaha in 1909. In recent years there have usually been about ten counting locations in Nebraska, with notably long sequences having developed in Lincoln, Omaha, and Scottsbluff and at Lake McConaughy. This last-named location invariably has the highest state count, often tallying about one hundred species, whereas species counts from Lincoln and Omaha are usually in the seventies. Since 1988 the Great Backyard Bird Count has attempted to measure February birds, with an emphasis on species likely to be seen in backyards and near homes. During the 2011 counts in Nebraska, 105 species and over three hundred thousand birds were counted.

Several years ago I decided that I could exploit the Christmas Count's vast amount of population data to test the idea that moderating winter climates in the Great Plains during the past half century may have resulted in a northward shift of early winter bird populations. I chose about two hundred bird species known to winter in the Great Plains and selected a forty-year period, from 1967–68 to 2006–7, for my study. As I was teaching an ornithology class at Cedar Point in the summer of 2008, I decided to make an analysis of these data a class project. I asked each of my eighteen students to analyze statistically the count data of eleven species, for every state from North Dakota through the Texas Panhandle, over the four decadal intervals.

After all these data had been accumulated, I began to summarize and analyze them (details can be found in Johnsgard and Shane 2009). The broad conclusion of the study was that a large proportion of the more than two hundred species analyzed exhibited a significant northward winter movement in their abundance peaks, which often shifted one state northward and sometimes moved even farther north. These results support the increasingly accepted position that significant climatic warming has been occurring in the Great Plains since at least the 1960s.

It is also of interest to consider the eighteen species of greatest average winter abundance in Nebraska, whose identities and numbers have never before been estimated. It should be remembered that most of the Christmas Counts are centered in cities (the count areas consist of circles fifteen miles in diameter), so the resulting numbers are probably biased toward urban and suburban birds, such as starlings and house sparrows. Some birders also choose to concentrate on watching bird feeders rather than doing outside surveys, which means that typical backyard birds are well represented. Lakes and reservoirs are also attractive counting sites, which results in large numbers of waterfowl and other aquatic birds being counted. Yet, because of the large number of counts and counters, a fair representation survey of a state's winter bird life is possible.

Typical aquatic birds that appear in the greatest numbers on Nebraska's Christmas Counts are the common mallard, Canada goose, and common merganser. All are large, cold-tolerant birds that will overwinter wherever open water can be found. The Canada goose has increased markedly in Nebraska over the time period studied, while the mallard has apparently shifted its major early winter population northward into South Dakota. Common mergansers have remained concentrated farther south in Oklahoma; a certainty of open water is necessary for these fish-eating ducks.

Red-winged blackbirds are seed- and grain-eating birds that concentrate in grain fields during fall but gravitate to marshlands during winter. They are one of the most abundant of all North American land birds, with a continental population estimated at more than two hundred million. European starlings are far more city-adapted than red-winged blackbirds, and in many cities they are easily the most common species year-round. In the same manner, the also-introduced house sparrow is most at home in cities but extends into the country around farms and homesteads. Both species are nonmigratory and appear to be in decline, at least in Nebraska.

American tree sparrows are remarkably cold tolerant considering their tiny size, with maximum early winter densities in Kansas. They tend to remain in the country, scratching on the ground for seeds in snow-free sites, but will rarely visit bird feeders. Their continental population has been estimated at about twenty-six million birds.

Horned larks are strictly open-country birds and in Nebraska are likely to be seen in snow-free grain fields, sometimes in large roaming flocks and frequently in the company of Lapland longspurs and snow buntings. None of these relatively abundant species is likely to be seen within the city limits of a town.

American robins are familiar backyard birds during summer, but as cold weather moves into Nebraska they flock up and move variably southward, depending on the intensity of the winter and the availability of fruits, their primary winter diet. The robin is the only species described here that has seemingly shifted its population density south in the past half century. This is probably a statistical anomaly, but wintering robins are highly nomadic, and flocks of as many as ten million birds have been reported in the southeastern states. Their continental population has been estimated as about 320 million.

Judging from the available Christmas Count data, Nebraska may be the best state in which to see Harris's sparrows during early winter. They migrate south from their arctic nesting areas in central Canada to winter in the central Great Plains, especially eastern Nebraska. They mostly are found in country habitats that offer weedy grasslands, brushlands, or open woodlands. They come to feeders only during times of severe weather, sometimes in the company of white-throated or white-crowed sparrows. American crows may also be found in habitats ranging from open country, which they scavenge widely for food, to fairly heady woods, where they roost, sometimes in very large flocks. In recent decades crows have become increasingly common in towns and cities, where they may look for suet in bird feeders but remain much more wary than woodpeckers or blue jays.

Except for the cedar waxwing, all of the remaining species judged as abundant Nebraska wintering birds are typical bird-feeder species. The cedar waxwing is a berry and fruit eater, especially favoring juniper (red cedar) berries. As such, cedar waxwings inhabit shelterbelts and other places where junipers are present and in towns seek out crab apples or other ornamental fruiting trees. Red cedars are abundant across the Great Plains and largely account for the species' winter abundance here.

In sequence of their seeming descending winter abundance in Nebraska, the remaining species are northern cardinal, dark-eyed junco, black-capped chickadee, and house finch. Of these, the house finch is a relative newcomer to eastern Nebraska, having invaded the state in the 1960s from the east, although a small population has long been resident in the Panhandle. It quickly became addicted to bird-feeding stations and has gradually tended to displace the more skittish house sparrow in abundance at backyard bird feeders. Northern cardinals have also expanded their range in Nebraska during the past half century, slowly moving farther north as winters have moderated in severity and becoming favorites of people willing to provide them with unlimited black-oil sunflower seeds.

The black-capped chickadee is perhaps the tamest and thus most visually appealing of all the bird-feeder species in Nebraska, although white-breasted and red-breasted nuthatches might win in the tameness category. Dark-eyed juncos are probably the most abundant wintering songbird in Nebraska and the entire Great Plains. Their continental abundance has been estimated at 260 million birds, easily edging out the cardinal's 100 million, the chickadee's 34 million, and the house finch's 21 million.

Over the forty years studied, ten of the eighteen species mentioned have exhibited apparent population increases in the final decade relative to their long-term Nebraska averages, three have apparently been stable, and five have shown apparent declines. Over the same period, twelve of these species have exhibited apparent northward statewide population shifts judging from average state density estimates, one species (the American robin) has seemingly moved south, and six have shown no statewide shifts.

Nebraska data from 2011 for the Great Backyard Bird Count provide a slightly different assessment of winter birds in the state. Based on the number of checklists reporting the species, the ten most commonly encountered birds were (in descending order) American robin, American goldfinch, downy woodpecker, house sparrow, blue jay, house finch, black-capped chickadee, Canada goose, European starling, and mourning dove. Surprisingly, the species seen in largest numbers was the snow goose, which is far from being a "backyard" bird but is often seen flying over towns in enormous flocks. It is a testimony to changing climates that the snow goose, long a symbol of early spring and late fall in the upper Great Plains, is now an abundant winter bird!

PART THREE

The View from a High Hill

Ferruginous hawk

21

A Summing Up

I began writing this final essay on Easter Sunday, the last day of March, 2013. I had just returned from a trip to the central Platte Valley for a celebration of spring and one last view of the sandhill cranes before they had all migrated north. Enduring several late winter snowstorms, they finally received the benefits of cloudless skies and a full moon to help them find the landmarks and northern wetlands that they would use for guidance and security on their remaining two-to-three-thousand-mile trips to their tundra breeding grounds. In spite of a cold spring, their departure was unusually early this year, perhaps because of depleted corn supplies in the face of competition for food with geese and a smaller than normal 2012 corn crop.

Of all these annual passages, seeing the cranes arrive and depart is always the most heart-tugging for me. They represent my deepest emotional connection to Nebraska and one of the primary reasons that I decided I wanted to spend the rest of my life here, less than a year after setting foot in the state for the first time in my life. In spite of having grown up less than three hundred miles north of Nebraska's northern border, I had never entered the state during my first thirty years of life and had acquired only two informal guidelines to help prepare me for what, as a biologist, I might find there. As an undergraduate student at North Dakota State University I had visited the Delta Waterfowl Research Station, west of Winnipeg, Manitoba, to observe its waterfowl research program and visit with its director, H. A. Hochbaum. In one of our discussions he mentioned that, in his view, the Nebraska Sandhills' wetlands were probably second only to North Dakota's prairie potholes as the greatest duck-production region south of Canada.

Later, while a first-year doctoral student at Cornell University, I listened with great interest to the experiences of two somewhat more advanced graduate student colleagues. They had engaged in summer field research along Nebraska's central Platte Valley, determining the amounts of range overlap and hybridization rates among several species pairs of songbirds and woodpeckers. According to them, the riparian forests of the central Platte Valley in May were an ornithologist's dream, teeming with fascinating and sometimes hybridizing species of orioles, buntings, towhees, and flickers.

Five years later, while finishing two years of postdoctoral research in England, those incidents were on my mind when I applied for, and happily accepted, a teaching position at the University of Nebraska, in Lincoln. Since then, as I have crossed and recrossed nearly every highway in the state (plus uncountable "unimproved" roads and a few unmarked Sandhills trails), I have come to love Nebraska's scenic and often pristine rivers, its tallgrass prairie remnants as well as its still-boundless Sandhills prairies, its myriad freshwater and alkaline wetlands, and especially its birds and other wildlife.

∾

During my roughly one-hundred-mile late-March drive to Grand Island to wish the cranes godspeed, I began to worry. I noticed there was still essentially no green vegetation visible from the interstate highway, except around some scattered patches of melting snow in shaded sites. The second year of an exceptionally severe drought had left Nebraska and most of the other Great Plains states with parched fields and mostly dry wetlands. Nevertheless, in 2012 Nebraska's cornfields produced an only slightly below average crop of about 1.3 billion bushels, thanks largely to more than one hundred thousand center-pivot irrigation systems and a still-bountiful and unpolluted

Ogallala Aquifer. Yet nearly all the corn stubble visible from the highway had been cut down at nearly ground level, removing all the aboveground biomass. The resulting nearly barren land offered virtually no food for wildlife, exposed the soil to potentially severe erosion, and promoted desiccation of the subsoil more severely than would have been the case if some vegetational cover had been left behind.

Nebraska is now the country's fourth-largest consumer of crop insurance, and 75 percent of the nearly $500 million paid out in federally funded indemnity payments in 2012 were the result of claims related to corn. South of the Platte Valley, the Rainwater Basin was nearly dry by March 2013. The subnormal snowfall of the previous winter had not replenished the region's wetlands, and the Weather Bureau's forecast for the summer months offered no basis for optimism. In association with the drought, during 2012 Nebraska also suffered one of its worst years in history as to wildfire damage, with almost five hundred thousand acres of grasslands and forests burned across the state. The spring of 2013 was marked by official estimates of extreme to exceptional drought conditions across nearly the entire state.

The years of bountiful annual rains and nearly unlimited extractions of water from both our surface-water supplies and the Ogallala Aquifer are evidently now behind us. In April 2013 Lake McConaughy was at only 66 percent of its capacity, and its spring inflow rates had been at about 75 percent of normal. The snowpack in the North Platte River Basin was then only 79 percent of average, and that of the South Platte River Basin was 83 percent of average. At that time the Central Nebraska Public Power and Irrigation District announced plans to restrict irrigation allotments for its customers to ten inches of water per acre, rather than the usual eighteen inches.

The world is now confronted with threats of global climate changes, foretelling even worse droughts, hotter summers, and international water and energy crises. Even though we know that many of the adverse effects of global warming are caused by the burning of fossil fuel, our state's senators and representatives are determined to risk Nebraska's ecological and economic future by allowing a gigantic oil pipeline to be built that will bisect central Nebraska from north to south. In spite of a minor face-saving shift in routing, it is still to be partly situated on a highly sandy base, and only a few feet above the top of our precious and still-unpolluted Ogallala Aquifer. It will cross the Niobrara, Loup, and Platte Rivers, passing through the breeding ranges of the endangered interior least tern and threatened piping plover, run closely parallel to the migration route of the endangered whooping crane, and transect the middle of the known Nebraska range of the endangered American burying beetle. Nebraska's most prominent politicians probably wouldn't know a least tern from a U-turn, but they can evidently detect the presence of potential money at great distances.

In spite of such ominous threats to Nebraska's advertised "good life," many welcome changes have occurred with regard to the state's environmental stewardship and its ecological monitoring. The annual Audubon Christmas Bird Counts, begun in 1900, were conducted in Lincoln and Omaha as early as 1909 and in Scottsbluff starting in 1949. Lincoln's Christmas Count database is now continuous for nearly sixty years and Scottsbluff's for more than fifty years. A total of about ten to fifteen Nebraska sites now are annually surveyed in Christmas Counts, providing a valuable index to long-term changes in our winter bird populations, as was described in chapter 20.

The value of the Platte as spring habitat for migrating sandhill and whooping cranes has attracted naturalists for several decades. The crane migration was initially celebrated in Grand Island by its Big Bend National Audubon Society chapter in 1970. The celebration was later moved to Kearney and until 2013 was called the Rivers and Wildlife Celebration. This annual celebration, now one of the longest-running in America, has attracted more than eight thousand people over the past four decades and provides a strong catalyst for environmental education and support for protecting the river, the cranes, and associated wildlife.

Thanks to a major bequest from a Massachusetts benefactor, in 1974 the National Audubon Society purchased a riverine strip consisting of about four shoreline miles and more than one thousand acres of mostly prime crane meadow habitat along the central Platte River near Gibbon. The sanctuary was named the Lillian Annette Rowe Sanctuary, in honor of the woman whose generous gift allowed for the area's purchase, at the very time that efforts by federal agencies to establish a national wildlife refuge in the region through land condemnation had failed miserably.

In 1999 Rowe Sanctuary began a capital campaign to add an education center, and the Iain Nicolson Audubon Center was completed in 2003 by a gift from a California woman, Margaret Nicolson, in memory of her late husband. Centered near one of the largest sandhill crane roost sites in the central Platte Valley, the center attracts about fifteen thousand people annually to see this amazing spectacle, adding to the local economy an estimated minimum of $20 million annually. In 2006 and 2008 an additional six hundred acres were added to Rowe Sanctuary's riparian wetlands, bringing its total to nineteen hundred acres by 2013. Audubon manages this riverine complex in a way that provides spring staging habitat for whooping and sandhill cranes, as well as nesting habitat for piping plovers and least terns, two other nationally endangered or threatened species.

Shortly after the passage of the 1972 Endangered Species Act, the central Platte Valley was identified as critical habitat for the nationally endangered whooping crane, one of America's rarest and most spectacular birds. On the basis of this critical habitat designation, the National Wildlife Federation and several other conservation groups mounted a legal challenge against the construction of Grayrocks Dam, being built on a tributary of the North Platte River in Wyoming. That 1978 lawsuit resulted in a $7.5 million mitigation fund. This fund has allowed for the establishment of the Platte River Whooping Crane Habitat Maintenance Trust (now the Crane Trust), headquartered near Grand Island, and the purchase and restoration of critical wet meadow and shoreline habitat on the Platte River between Grand Island and Kearney. Associated lands purchased later by the trust and other conservation groups have helped secure the most valuable remaining wetlands and riverine roosting areas. In 2013 the trust assumed management of the Nebraska Nature and Visitor Center (now the Crane Trust Nature and Visitor Center) at the Alda I-80 interchange near Grand Island, as part of an effort to perform an expanded role in promoting regional environmental education and undertaking ecological research in the Platte Valley, especially on cranes.

In 1994 the U.S. Fish and Wildlife Service decided to impose restrictions on central Platte water use to protect the four threatened or endangered species using the central and lower Platte Valley. The restrictions required all users, primarily irrigators, to maintain sufficient water flows to protect these species. As a result, all interested parties, including the Audubon Society, began to negotiate means of achieving these ends. In 1997 a Platte River Cooperative Agreement was initially agreed upon, although it took another decade to iron out all the details.

In 2003 Nebraska's irrigators challenged the cooperative agreement by contending that the Platte River (which they had largely dewatered and degraded) no longer constituted critical habitat for whooping cranes and that its water management should not be subjected to the restrictions imposed by the Endangered Species Act. As a result, the National Academy of Sciences was asked to evaluate this question. The academy confirmed that the central Platte River still represented critical habitat for whooping cranes and affected the status of three other threatened or endangered species and that the cooperative agreement satisfies the requirements of the Endangered Species Act. That judgment set the stage for an eventual thirteen-year, multistate agreement on the Platte's habitat restoration and water conservation, which was approved in 2008 by Colorado, Wyoming, Nebraska, and the federal government. This plan allowed, in part, for an expenditure of more than $300 million for the purchase, easement acquisition, and improvement of central Platte wetland habitats and has already been of immeasurable importance to cranes, waterfowl, and other wetland-dependent wildlife.

In 1980, as part of its mission to protect ecologically important lands and waters, the Nature Conservancy purchased fifty-two thousand acres representing about twenty-five riverfront miles on the river's south side and about eight miles along its north side. In 1985 Senator James Exon introduced a bill to designate a seventy-nine-mile stretch of the Niobrara River Valley as a Nationally Scenic River, an area that centered on the Nature Conservancy's Niobrara Valley Preserve. After a contentious legislative battle, the act was approved in 1991, and much of the Niobrara's most scenic reaches are now protected from development and freely open to the public for canoeing, rafting, and other nonconsumptive recreation.

In 2002, 218 acres of Niobrara River shoreline were acquired by the National Audubon Society and turned over to the Nebraska Game and Parks Commission to form the Fred Thomas Wildlife Management Area, commemorating a well-known *Omaha World-Herald* reporter and longtime advocate of protecting the Niobrara Valley. In the following year, Audubon of Kansas acquired ownership of a five-thousand-acre ranch, the Hutton Niobrara Ranch Wildlife Sanctuary in the central Niobrara Valley, which is being developed as an important site for wildlife conservation, environmental education, and nature appreciation.

Conservation easements coordinated by the Nebraska Land Trust and funded in part by the Nebraska Environmental Trust have also helped to preserve the integrity of the Niobrara Valley. Nevertheless, conservationists have to remain vigilant in protecting the Niobrara Valley from constant political pressures to commercialize this ecologically important region and from other threats to its vulnerable resources. During the record-setting torrid summer of 2012 some thirty thousand acres of the Niobrara Valley Preserve burned as a result of lightning-caused wildfires, producing enormous changes in the forest and grassland vegetation that will persist for many decades. Similar massive forest fires also occurred in the Pine Ridge region, the 2012 fires in total affecting more than 270,000 acres, extending from the central Niobrara Valley west almost to the Wyoming border.

Since 1961 several bird species that have long been absent as breeders in the state have begun to nest again in Nebraska. As noted in chapter 13, with the aid of hacking and nest-site preparation efforts, peregrine falcons have been nesting regularly in Omaha since 1998 and in Lincoln since 2003. Bald eagles returned to attempt nesting in the state within a year after DDT was outlawed in the early 1970s, although the first successful known nesting did not occur until 1991. Bald eagle nests have since been reported in at least sixty-two counties, mostly along wooded river valleys. After a hiatus of nearly a century, greater sandhill cranes have again nested in the state at least intermittently since 1999. Whooping cranes have also responded to habitat improvements along the central Platte River and are now again regularly roosting there during their spring migrations.

In west central Nebraska, nesting by black-necked stilts was reported for the first time in the state at Crescent Lake National Wildlife Refuge during the 1980s, and nestings have since occurred elsewhere in the Sandhills. Marbled godwits have also been found nesting sparingly in the northern Sandhills. White-faced ibises have similarly expanded during the past decade from a few known nesting sites in the Sandhills and Rainwater Basin, and the very similar but more southern-oriented glossy ibis has been increasingly seen and has possibly also nested recently in the state. Likewise, since 2008 ospreys have made nesting efforts on four different nesting platforms in Keith and Scotts Bluff Counties.

Some eastern bird species have moved westward along Nebraska's increasingly forested rivers, such as the wood duck, red-bellied woodpecker, and Baltimore oriole. Eastern Nebraska's tiny breeding populations of several deciduous forest-dependent birds, such as the pileated woodpecker, Kentucky warbler, and summer tanager, also seem to be slowly expanding their breeding ranges in the mature forests of the Missouri Valley.

A few western-oriented songbirds have recently moved eastward into the scrublands of Nebraska's Panhandle from Wyoming and Colorado in recent years, such as the western race of the blue-gray gnatcatcher and the cordilleran flycatcher. Given the prospects of more frequent and more severe future droughts, this west-to-east movement by arid-adapted species is likely to increase. In the winter of 2012–13 there was a major influx of Rocky Mountain finches, crossbills, and evening grosbeaks into Nebraska, probably reflecting poor conifer cone production in the drought-impacted Wyoming mountains.

Several species have expanded their breeding ranges northward into Nebraska since the 1960s, such as the chuck-wills-widow, scissor-tailed flycatcher, great-tailed grackle, and Mississippi kite. Mississippi kites have nested in several Nebraska towns since 1991, including Ogallala, Red Cloud, Benkleman, and Imperial, as part of an apparent northern range expansion. The southern gray-headed race (*J. h. dorsalis*) of the dark-eyed junco is becoming almost regular in western Nebraska, and some distinctly southern birds, such as the crested caracara, Harris's hawk, zone-tailed hawk, and hooded oriole, have put in surprise recent appearances.

One important development reflecting cultural changes in attitudes toward wildlife appreciation over the past half century has been the Nebraska Game and Park Commission's broadened viewpoint as to its statewide mission. In the early 1960s the agency was largely preoccupied with promoting hunting, fishing, and enjoyment of our state parks and other recreational areas. By the start of the twenty-first century it had hired its first nongame biologist, had helped the Nebraska Ornithologists' Union publish the results of the state's first breeding bird survey, and had spearheaded the formation of the Nebraska Partnership for All-Bird Conservation (now the Nebraska Bird Partnership, www.nebraskabirds.org). This loose affiliation of conservation-oriented organizations and agencies has energized a cooperative enthusiasm that has produced an online version of my *Nebraska Bird-Finding Guide* (www.nebraskabirding trails.com), developed an interactive curriculum for bird study directed toward children in the fifth to eighth grades (Project Beak, http://projectbeak.org/), and generated an online bird identification guide (the Nebraska Bird Library, www.nebraskabirdlibrary .org), designed to help anybody identify the more than 450 species that occur in the state.

Also of great importance to the preservation of Nebraska's natural heritage were the passage of Nebraska's Threatened and Endangered Species Act and the formation of the Nebraska Natural Legacy Project (http://outdoornebraska.gov/), which works to focus conservation efforts on native flora and fauna in the state's forty biologically unique landscapes.

From a landscape conservation perspective, dozens of new state-owned wildlife management areas and state recreation areas have been established through the efforts of Nebraska's Game and Park Commission since 1961, bringing the current total of state parks and state historical parks to 19, state recreation areas to 60, and wildlife management areas to 283. Several state parks that include important or unusual habitats have also been established in the past few decades, including Bowring Ranch State Historical Park in 1985, Eugene T. Mahoney State Park in 1991, Ashfall Fossil Beds State Historical Park in 1991, and Smith Falls State Park in 1992. Interpretive centers have been constructed in the Wildcat Hills State Recreation Area, Ash Hollow State Recreation Area, and, in cooperation with the University of Nebraska State Museum, Ashfall Fossil Beds State Historical Park.

Lincoln's Wachiska chapter of the National Audubon Society was established in 1973 and celebrated forty years of conservation work in 2013. In 1976 it began what would become a statewide raptor rehabilitation program, Raptor Recovery Nebraska. During its first thirty-five years of operation the center treated over fifty-three hundred hawks, eagles, and vultures and nearly five thousand owls, with a return-to-the-wild rate of nearly 50 percent. The treated birds have included all of Nebraska's eighteen species of hawks, eagles, and falcons, its nine species of owls, and the turkey vulture. In 2013 the center was adopted by the Fontenelle Forest Nature Center in Omaha, helping to strengthen its long-term prospects.

In 1978 Wachiska leased 240 acres of what was once eight hundred acres of historically important tallgrass prairie near Lincoln (Nine-Mile Prairie) from the Lincoln Airport Authority. This effort began a process that eventually resulted in the prairie's protection through its acquisition by the University of Nebraska Foundation. This action also heralded the start of an ongoing program of Wachiska to preserve tallgrass prairies throughout seventeen southeastern Nebraska counties, either by purchase or through conservation easements. The first of these prairies, the five-acre Henry Wulf Tallgrass Prairie near Lincoln, was protected by a conservation easement in 1994. Since then, twenty-one additional prairies in the seventeen-county region have been protected under conservation easements. Five prairies, totaling more than eighty acres, have been acquired by Wachiska, as has a four-hundred-acre farm that is gradually being reverted to prairie. Efforts are underway by the Wachiska chapter to provide educational opportunities for schoolchildren on all these prairies, which sometimes support as many as three hundred or more plant species and hundreds of species of invertebrates and vertebrates.

A major step in prairie conservation and education was made in 1998, when Nebraska's Audubon Society purchased the 610-acre Kathy O'Brien Ranch in southern Lancaster County. This rare tract of virgin tallgrass prairie, now named Spring Creek Prairie Audubon Center, was expanded by 26 acres in 2000, by 16 more in 2003, and in 2007 an adjoining quarter-section was obtained, bringing the prairie's total acreage to 808. Nearly 50 more acres were added in 2013. Nearby prairies under conservation easements bring the overall area of locally protected grasslands to nearly 20,000 acres.

In 2005 construction began on a new education center at Spring Creek. This building opened in 2006, and within a few years was attracting upwards of ten thousand people yearly. As described in chapter 6, in 2009 Spring Creek Prairie began a ten-year Prairie Immersion Project, during which half of all the several thousand fourth-grade students in the Lincoln Public Schools visit Spring Creek Prairie. The others visit a similar relict prairie at Lincoln's Pioneers Park Nature Center, which celebrated fifty years of nature education in 2013.

One of the most significant events in the protection of Nebraska's natural environments during the past half century was the establishment in 1992 of the Nebraska Environmental Trust, using proceeds from state gambling profits. The trust's stated mission is " to conserve, enhance and restore the natural environments of Nebraska." During its first decade of existence, and in spite of regular efforts by special-interest groups to divert its funds to other purposes, the trust distributed about $40 million to environmental causes. Who said gambling is all bad?

ॐ

The wintering bald eagles, snow geese, and other arctic-bound geese had mostly left Nebraska by the end of March 2013, the geese tracing fine, wavering lines across the sky almost throughout the day or revealing their passage from elevations too high for my slowly weakening eyes to detect them. In mid-April I was viewing greater prairie-chickens and sharp-tailed grouse performing their spectacular displays on their "lekking" grounds in the Nebraska Sandhills, and by early May I was observing migrant shorebirds and songbirds. Where else but in Nebraska could a naturalist find a more fulfilling life?

At one time early in my University of Nebraska career I thought that, if I lived that long, I might learn enough about the state's ecology, plants, and wildlife to write a book about a few of them. A half century later, I now know that I will never be able to describe all of the great stories that could be discovered among Nebraska's biological attractions. Those stories that are told on the preceding pages reflect only a few facets of the state's hidden natural treasures, as seen by a single person; it remains for others to pursue their own dreams and to pass on their knowledge and achievements to still another generation. It should be clear from this brief review of what has transpired during Nebraska's last half century that there are hosts of Nebraskans willing to continue our collective goals of resource conservation and environmental appreciation well into the future.

APPENDIX

Latin Names of Plants and Animals Mentioned in the Text

Lark sparrow and eastern red cedar

American avocet, *Recurvirostra americana*
American bellflower, *Campanula americana*
American bittern, *Botaurus lentiginosus*
American coot, *Fulica americana*
American crow, *Corvus brachyrhynchos*
American golden-plover, *Pluvialis dominica*
American goldfinch, *Carduelis tristis*
American kestrel, *Falco sparverius*
American linden, *Tilia americana*

American plum, *Prunus americana*
American robin, *Turdus migratorius*
American tree sparrow, *Spizella arborea*
American white pelican, *Pelecanus erythrorhynchos*
American wigeon, *Anas americana*
"antelope" (pronghorn), *Antilocapra americana*
arctic fox, *Alopex lagopus*
badger, *Taxidea taxus*
Baird's sandpiper, *Calidris bairdii*
bald eagle, *Haliaeetus leucocephalus*
Baltimore oriole, *Icterus galbula*
barn owl, *Tyto alba*
barn swallow, *Riparia riparia*
barred owl, *Strix varia*
Barrow's goldeneye, *Bucephala islandica*
beaver, *Castor canadensis*
Belted kingfisher, *Ceryle alcyon*
bergamot, *Monarda fistulosa*
Bewick's swan, *Cygnus bewickii*
big bluestem, *Andropogon gerardii*
bison, *Bison bison*
bitternut hickory, *Carya cordiformis*
black-billed cuckoo, *Coccyzus erythropthalmus*
black-billed magpie, *Pica hudsonia*
black-capped chickadee, *Poecile atricapillus*
black-crowned night heron, *Nycticorax nycticorax*
black-footed ferret, *Mustela nigripes*
black grouse, *Tetrao tetrix*
black-headed grosbeak, *Pheucticus melanocephalus*
black-headed gull, *Larus ridibundus*
black-necked stilt, *Himantopus mexicanus*
black-tailed prairie dog, *Cynomys ludovicianus*
black tern, *Chlidonias niger*
black walnut, *Juglans nigra*
blue-gray gnatcatcher, *P. c. amnoenissima*

blue jay, *Cyanocitta cristata*
blue spiderwort, *Tradescantia bractiata*
blue violet, *Viola sororia*
blue-winged teal, *Anas discors*
bobolink, *Dolichonyx oryzivorus*
bogus yucca moths, *Prodoxus* spp.
Bohemian waxwing, *Bombycilla garrulus*
bristly greenbrier, *Smilax hispida*
broad-tailed hummingbird, *Selasphorus platycercus*
brown-headed cowbird, *Molothrus ater*
brown pelican, *Pelecanus occidentalis*
brown thrasher, *Toxostoma rufum*
buffalo grass, *Buchloe dactyloides*
buff-breasted sandpiper, *Tryngites subruficollis*
bufflehead, *Bucephala albeola*
bullheads, *Ictalurus* spp.
Bullock's oriole, *Icterus bullockii*
bur oak, *Quercus macrocarpa*
burrowing owl, *Athene cunicularia*
cackling goose, *Brant hutchinsii*
calliope hummingbird, *Stellula calliope*
Canada goose, *Branta canadensis*
canvasback, *Aythya valisineria*
cardinal flower, *Lobelia cardinalis*
caribou, *Rangifer tarandus*
carp, *Cyprinis carpio*
carrion flower, *Smilax lasioneura*
cedar waxwing, *Bombycilla cedrorum*
chipping sparrow, *Spizella passerina*
chuck-wills-widow, *Caprimulgus carolinensis*
cinnamon teal *Anas cyanoptera*
Clark's grebe, *Aechmophorus clarkii*
cliff swallow, *Petrochelidon pyrrhonota*
columbine, *Aquilegia canadensis*
common (Old World) cuckoo, *Cuculus canorus*

common crane, *Grus grus*
common goldeneye, *Bucephala clangula*
common merganser, *Mergus merganser*
common nighthawk, *Chordeiles minor*
common poorwill, *Phalaenoptilus nuttallii*
common yellowthroat, *Geothlypis trichas*
coneflowers, *Ratibida* spp.
Cooper's hawk, *Accipiter cooperii*
cordilleran flycatcher, *Empidonax occidentalis*
coyote, *Canis latrans*
crab apple, *Pyrus ioensis*
crayfish, *Cambaridae* spp.
caracara, *Caracara cheriway*
cutthroat trout, *Salmo clarkii*
daisy fleabane, *Erigeron annus*
dark-eyed junco, *Junco hyemalis*
deer (white-tailed), *Odocoileus virginianus*
dickcissel, *Spiza americana*
double-crested cormorant, *Phalacrocorax auritus*
downy gentian, *Gentiana puberulenta*
downy woodpecker, *Picoides pubescens*
Dutchman's-breeches, *Dicentra cucullaria*
eared grebe, *Podiceps nigricollis*
eastern fox squirrel *Sciurus niger*
eastern meadowlark, *Sturnella magna*
eastern phoebe, *Sayornis phoebe*
eastern screech-owl, *Otus asio*
eastern towhee, *Pipilo erythrophthalmus*
eastern virgin's bower, *Clematis ligusticifolia*
European starling, *Sturnus vulgaris*
ferruginous hawk, *Buteo regalis*
field sparrow, *Spizella pusilla*
Forster's tern, *Sterna forsteri*
gadwall, *Anas strepera*
gayfeathers, *Liatris* spp.

glossy ibis, *Plegadis falcinellus*
golden eagle, *Aquila chrysaetos*
goldenrods, *Solidago* spp.
grasshopper sparrow, *Ammodramus savannarum*
gray-crowned rosy finch, *Leucosticte tephrocotis*
gray wolf, *Canis lupus*
great black-backed gull, *Larus marinus*
greater prairie-chicken, *Tympanuchus cupido*
greater sage-grouse, *Centrocercus urophasianus*
greater white-fronted goose, *Anser albifrons*
great gray owl, *Strix nebulosa*
great horned owl, *Bubo virginianus*
great-tailed grackle, *Quiscalus major*
green-winged teal, *Anas crecca*
ground squirrel, *Spermophilus* spp.
hairy grama, *Bouteloua hirsuta*
Harlan's hawk, *Buteo jamaicensis harlani*
Harris's hawk, *Parabuteo unicinctus*
Harris's sparrow, *Zonotrichia querula*
Henslow's sparrow, *Ammodramus henslowii*
hoary puccoon, *Lithospermum canescens*
hooded merganser, *Lophodytes cucullatus*
hooded oriole, *Icterus cucullatus*
horned lark, *Eremophila alpestris*
house finch, *Carpodacus mexicanus*
house sparrow, *Passer domesticus*
house wren, *Troglodytes aedon*
Hudsonian godwit, *Limosa haemastica*
Iceland gull, *Larus glaucoides*
Inca dove, *Columbina inca*
Indiangrass, *Sorghastrum nutans*
indigo bunting, *Passerina cyanea*
jack-in-the-pulpit, *Arisaema triphyllum*
jackrabbits, *Lepus* spp.
jaeger, *Stercorarius* spp.

Junegrass, *Koeleria macrantha*
juniper (red cedar), *Juniperus virginiana*
Kentucky warbler, *Geothlypis formosa*
killdeer, *Charadrius vociferus*
Lapland longspur, *Calcarius lapponicus*
lark sparrow, *Chondestes grammacus*
lazuli bunting, *Passerina amoena*
least sandpiper, *Calidris minutilla*
least tern, *Sternula antillarum*
lemming, *Lemmus* and *Dicrostonyx* spp.
lesser scaup, *Aythya affinis*
little bluestem, *Schizachyrium scoparium*
little gull, *Hydrocoleus minutus*
long-billed curlew, *Numenius americanus*
long-billed dowitcher, *Limnodromus scolopaceus*
long-eared owl, *Asio otus*
mallard, *Anas platyrhynchos*
marbled godwit, *Limosa fedoa*
marsh wren, *Cistothorus palustris*
mew gull, *Larus canus*
milkweeds, *Asclepias* spp.
Mississippi kite, *Ictinia mississippiensis*
mountain chickadee, *Poecile gambeli*
mourning dove, *Zenaida macroura*
muskrat, *Ondatra zibethicus*
mute swan, *Cygnus olor*
New England aster, *Aster novae-angliae*
northern bobwhite, *Colinus virginianus*
northern cardinal, *Cardinalis cardinalis*
northern flicker, *Colaptes auratus*
northern goshawk, *Accipiter gentilis*
northern harrier, *Circus cyaneus*
northern pintail, *Anas acuta*
northern saw-whet owl, *Aegolius acadicus*
northern shoveler, *Anas clypeata*

orchard oriole, *Icterus spurius*
osprey, *Pandion haliaetus*
Pacific loon, *Gavia pacifica*
pale touch-me-not, *Impatiens pallida*
pallid sturgeon, *Scaphirhyncus albus*
peregrine falcon, *Falco peregrinus*
pileated woodpecker, *Dryocopus pileatus*
piping plover, *Charadrius melodus*
prairie falcon, *Falco mexicanus*
prairie phlox, *Phlox pilosa*
prairie rose, *Rosa arkansana*
purple coneflower, *Ratibida columnifera*
raccoon, *Procyon lotor*
red-bellied woodpecker, *Melanerpes carolinus*
red-breasted nuthatch, *Sitta canadensis*
red fox, *Vulpes vulpes*
redhead, *Aythya americana*
red knot, *Calidris canutus*
red-necked grebe, *Podiceps grisegena*
red-necked phalarope, *Phalaropus lobatus*
red-tailed hawk, *Buteo jamaicensis*
red-winged blackbird, *Agelaius phoeniceus*
ring-necked duck, *Aythya collaris*
rock pigeon, *Columba livia*
rock wren, *Salpinctes obsoletus*
rose-breasted grosbeak, *Pheucticus ludovicianus*
Ross's goose, *Chen rossii*
rough-legged hawk, *Buteo lagopus*
ruby-throated hummingbird, *Archilochus colubris*
ruddy duck, *Oxyura jamaicensis*
ruddy turnstone, *Arenaria interpres*
rufous hummingbird, *Selasphorus rufus*
salvia, *Salvia* spp.
sand bluestem, *Andropogon hallii*
sand dropseed, *Sporobolus cryptandrus*

sanderling, *Calidris alba*
sandhill crane, *Grus canadensis*
sandreed, *Calamovilfa longifolia*
scissor-tailed flycatcher, *Tyrannus forficatus*
sedge wren, *Cistothorus platensis*
semipalmated sandpiper, *Calidris pusilla*
sharp-shinned hawk, *Accipiter striatus*
sharp-tailed grouse, *Tympanuchus phasianellus*
short-eared owl, *Asio flammeus*
snakeweed, *Gutierrezia sarothrae*
snow bunting, *Plectrophenax nivalis*
snow goose, *Chen caerulescens*
snowy owl, *Bubo scandiaca*
sora, *Porzana carolina*
spotted towhee, *Pipilo maculatus*
striped skunk, *Mephitis mephitis*
suckers, *Carpoides* and *Catastomus*
summer tanager, *Piranga rubra*
sunflowers, *Helianthus* spp.
swift fox, *Vulpes velox*
tall blazing star, *Liatris pycnostachia*
tiger salamander, *Ambystoma mavortium*
Trumpet-creeper, *Campsis radicans*
trumpeter swan, *Cygnus buccinator*
tufted duck, *Aythya fuligula*
tundra swan, *Cygnus columbianus*
turkey vulture, *Cathartes aura*
upland sandpiper, *Bartramia longicauda*
Virginia creeper, *Parthenocissus quinquefolia*
Virginia rail, *Rus limicola*
weasel, *Mustela* spp.
western grebe, *Aechmophorus occidentalis*
western meadowlark, *Sturnella neglecta*
western sandpiper, *Calidris mauri*
white-breasted nuthatch, *Sitta carolinensis*

white-crowned sparrow, *Zonotrichia leucophrys*
white-faced ibis, *Plegadis chihi*
white-rumped sandpiper, *Calidris fuscicollis*
white snakeroot, *Ageratina altissima*
white-throated sparrow, *Zonotrichia albicollis*
white-winged scoter, *Melanitta fusca*
whooper swan, *Cygnus cygnus*
whooping crane, *Grus americana*
wildebeest (bridled gnu), *Connochaetes taurinis*
willet, *Tringa semipalmatus*
willow, *Salix* spp.
Wilson's phalarope, *Phalaropus tricolor*
Wilson's snipe, *Gallinago delicata*
wood duck, *Aix sponsa*
yellow-billed cuckoo, *Coccyzus americanus*
yellow-breasted chat, *Icteria virens*
yellow warbler, *Setophaga petechia*
yucca (soapweed), *Yucca glauca*
yucca moth, *Tegeticula yuccasella*
zone-tailed hawk, *Buteo albonotatus*

BIBLIOGRAPHIC SOURCES

Hooded merganser

A Place Called Pahaku

Bristow. D. 2005. "Pahaku—Nebraska's Center of the World." *Nebraska Life* 9(6): 31–36.

Cunningham, D. 1985. "Pahuk Place." *Nebraskaland* 67(6): 27–31.

Harrison, A. T. 1984. "Pahuk Bluff Historic Natural Area: A Report on the Natural History, Ecology and Anthropological Values, with a Biological Inventory Survey, Master Plan and Management Recommendations." Typescript ms.; copy filed in Nebraska State Historic Society Library, Lincoln.

Johnsgard, P. A. 2010. "A Place Called Pahaku." *Prairie Fire* 4(6): 1, 19, 20, 23. http://www.prairiefirenewspaper.com/2010/06/a-place-called-pahaku.

Life and Hard Times of the Platte

Brown, M. B., and P. A. Johnsgard. 2013, *Birds of the Central Platte River Valley and Adjacent Counties*. Lincoln: Zea E-Books and Digital Commons, University of Nebraska–Lincoln. http://digitalcommons.unl.edu/zeabook/.

Freeman, D. M. 2010. *Implementing the Endangered Species Act on the Platte Basin Water Commons*. Boulder: University of Colorado Press.

Johnsgard, P. A. 2007. *A Guide to the Natural History of the Central Platte Valley of Nebraska*. Digital Commons, University of Nebraska–Lincoln. http://digitalcommons.unl.edu/biosciornithology/40.

———. 2008. *The Platte: Channels in Time*. 2nd ed. Lincoln: University of Nebraska Press.

———. 2012. *Nebraska's Wetlands: Their Wildlife and Ecology*. Nebraska Water Survey Paper No. 78. Lincoln: Conservation and Survey Division, Institute of Agriculture and Natural Resources, University of Nebraska–Lincoln.

Nebraska's Magical Spring Migration

Johnsgard, P. A. 1983. "The Platte: A River of Birds." *Nature Conservancy News*, 33(5): 6–15.

———. 2003. "Great Gathering on the Great Plains." *National Wildlife* 41(3): 20–29. Digital Commons, University of Nebraska–Lincoln. http://digitalcommons.unl.edu/johnsgard/38/.

———. 2009. "The Wings of March." *Prairie Fire* 3(3): 1, 17, 18, 19. http://www.prairiefirenewspaper.com/2009/03/nature-notes-wings-of-march.

———. 2012. *Wings over the Great Plains: Bird Migrations in the Central Flyway*. Lincoln: Zea E-Books and Digital Commons, University of Nebraska–Lincoln. http://digitalcommons.unl.edu/zeabook/13/. Also available as a paperback from Lulu Enterprises, http://www.lulu.com/spotlight/unllib.

Melema, B. 2013. "The Crane Trust: Sinking Roots, Deepening the Vision." *Nebraskaland* 91(2):32–35.

The Birds of Nebraska's Boondocks

Brown, C. R., and M. B. Brown. 2001. *Birds of the Cedar Point Biological Station*. Occasional Papers of the Cedar Point Biological Station No. 1. Lincoln NE: Cedar Point Biological Station.

Brown, M. B., S. Dinsmore and C. R. Brown. 2011. *Birds of Southwestern Nebraska*. Lincoln: Conservation and Survey Division, University of Nebraska.

Johnsgard, P. A. 2005. *A Nebraska Bird-Finding Guide*. Lincoln: Digital Commons, University of Nebraska–Lincoln. http://digitalcommons.unl.edu/biosciornithology/51. Also available in hardcover from Lulu Enterprises, http://lulu.com/spotlight/unllib.

———. 2012. "The Birds of Nebraska's Boondocks." *Prairie Fire* 6(4): 8–10. http://www.prairiefirenewspaper.com/2012/04/the-birds-of-nebraskas -boondocks.

———. 2013. *The Birds of Nebraska.* Rev. ed. Lincoln: Digital Commons, University of Nebraska–Lincoln. http://digitalcommons.unl.edu /biosciornithology/38.

What Is a Tallgrass Prairie?

Johnsgard, P. A. 2008. *A Guide to the Tallgrass Prairies of Eastern Nebraska and Adjacent States.* Digital Commons, University of Nebraska–Lincoln. http:// digitalcommons.unl.edu/biosciornithology/39.

———. 2009. "Autumn on the Prairie: Nebraska's Grasses." *Nebraska Life* 13(5): 18–21. Digital Commons, University of Nebraska–Lincoln. http:// digitalcommons.unl.edu/johnsgard/40/.

———. 2009. "Forbs and Grasses and Cheshire Cats: What Is a Tallgrass Prairie?" *Prairie Fire* 3(11): 3, 9. http://www.prairiefirenewspaper.com/2009/12 /forbs-and-grasses-and-cheshire-cats-what-is-a-tallgrass-prairie.

Spring Creek Audubon Prairie

Johnsgard, P. A. 2012. "Spring Creek Prairie Audubon Center: An 800-Acre Schoolhouse." *Prairie Fire* 6(10): 18–20, 22. http://www.prairiefirenews paper.com/2012/10/spring-creek-prairie-audubon-center-an-800-acre -schoolhouse/.

Williams, T. 2010. "Splendor in the Grass." *Audubon* 112(5): 52–55.

Snow Geese of the Central Flyway

Johnsgard, P. A. 1975. "The Lesser Snow Geese of Central North America. *Wildlife* 75: 63–68.

———. 1975. *Waterfowl of North America.* Lincoln: University of Nebraska Press. Rev. ed., 2010, Lincoln: Digital Commons, University of Nebraska– Lincoln. http://digitalcommons.unl.edu/biosciwaterfowlna/1.

———. 2010. "Snow Geese on the Great Plains." *Prairie Fire* 4(2): 12–15. http://www.prairiefirenewspaper.com/2010/2/snow-geese-on-the-great -plains.

A Congruence of Cranes

Johnsgard, P. A. 1991. *Crane Music: A Natural History of American Cranes.* Washington DC: Smithsonian Institution Press. Reprint, 1997, Lincoln: University of Nebraska Press.

———. 2003. "Nebraska's Sandhill Crane Populations: Past, Present and Future. *Nebraska Bird Review* 71: 175–78. http://digitalcommons.unl.edu /biosciornithology/16.

———. 2011. *The Sandhill and Whooping Cranes: Ancient Voices over America's Wetlands*. Lincoln: University of Nebraska Press,

Johnsgard, P. A., and K. Gil. 2011. "Sandhill Cranes: Nebraska's Avian Ambassadors-at-Large. *Prairie Fire* 5(3): 14, 15, 20. http://www.prairiefire newspaper.com/2011/02/sandhill-cranes-our-avian-ambassadors-at-large.

Whooping Cranes Are Still Surviving

Austin, J., and A. L. Rickert. 2001. *A Comprehensive Review of Observational Data and Site Evaluation Date of Migrant Whooping Cranes in the United States, 1943–1990*. Jamestown ND: Northern Prairie Research Center. http:// www.npwrc.usgs.gov/resource/birds/wcdata/index.htm.

Gil-Weir, K., F. Chavez-Ramirez, B. W. Johns, L. Craig-Moore, T. Stehn, and R. Silva. 2011. "Historical Breeding, Stopover and Wintering Distributions of a Whooping Crane Family." Abstracts of the 11th North American Crane Workshop and 34th Annual Meeting of the Waterbird Society, March 13–26, 2011, Grand Island NE, p. 36.

Gil-Weir, K., and P. A. Johnsgard. 2010. "The Whooping Cranes: Survivors against All Odds. *Prairie Fire* 4(9): 12, 13, 16, 22. http://www.prairiefire newspaper.com/2010/09/the-whooping-cranes-survivors-against-all-odds.

Harner, M. 2013. "Studying the Migration Patterns and Stopover Habitats of the Whooping Crane. *Prairie Fire* 7(4): 12–13.

Hartsup, B. 2012. "Whooping Cranes of the 60th Parallel." *Bugle* 38(1): 1–2.

Johnsgard, P. A. 1991. *Crane Music: A Natural History of American Cranes*. Washington DC: Smithsonian Institution Press. Reprint, 1997, Lincoln: University of Nebraska Press.

———. 2008. *The Status of Cranes of the World in 2008: A Supplement to* Crane Music. Lincoln: Digital Commons, University of Nebraska–Lincoln. http:// digitalcommons.unl.edu/biosciornithology/45/.

———. 2011. *The Sandhill and Whooping Cranes: Ancient Voices over America's Wetlands*. Lincoln: University of Nebraska Press.

———. 2012. *Wetland Birds of the Central Plains: South Dakota, Nebraska and Kansas*. Lincoln: Digital Commons, University of Nebraska–Lincoln. http://digitalcommons.unl.edu/zeabook/8. Also available in hardcopy from Zea E-Books, Lulu Enterprises, http://www.lulu.com/spotlight /unllib.

Williams, T. 2013. "Taking a Stand." *Audubon* 115(4): 26–34.

Strange Courtship of Prairie Grouse

Johnsgard, P. A. 1973. *Grouse and Quails of North America*. Lincoln: University of Nebraska Press. Also available at Digital Commons, University of Nebraska–Lincoln, http://digitalcommons.unl.edu/bioscigrouse/1.

————. 2002. *Grassland Grouse and Their Conservation.* Washington DC: Smithsonian Institution Press.

————. 2007. "A Dozen Squaretails and a Sharpy." *Nebraska Life* 11(2): 80–86.

————. 2010. "The Drums of April." *Prairie Fire* 4(4): 12–13. http://www.prairie firenewspaper.com/2010/04/the-drums-of-april.

————. 2013. "The Greater Prairie-Chicken: Spirit of the Tallgrass Prairie." *Prairie Fire* 7(4): 14–15. http://www.prairiefirenewspaper.com/2013/04/the -greater-prairie-chicken-spirit-of-the-tallgrass-prairie.

————. 2013. "The Grouse with the Pointed Tail." *Prairie Fire* 7(4): 16–18. http:// www.prairiefirenewspaper.com/2013/04/the-grouse-with-the-pointed-tail.

Shorebirds and Their Amazing Migrations

Johnsgard, P. A. 2011. "The Secretive Shorebirds: Nebraska's Phantom Migrants." *Prairie Fire* 5(4): 12–13. http://www.prairiefirenewspaper.com /2011/04/the-secretive-shorebirds-nebraskas-phantom-migrants.

Jorgensen, J. 2004. *An Overview of Shorebird Migration in the Eastern Rainwater Basin.* Occasional Paper No. 8. Lincoln: Nebraska Ornithologists' Union.

Jorgensen, J. G. 2012. *Birds of the Rainwater Basin, Nebraska.* Lincoln: Nebraska Game and Parks Commission. http://outdoornebraska.ne.gov/wildlife /programs/nongame/NGBirds/pdf/Birds%20of%20the%20rainwater% 20basin%20version%201.0%20(May%202012).pdf.

Birds of the Tallgrass Prairie

Johnsgard, P. A. 2001. *Prairie Birds: Fragile Splendor in the Great Plains.* Lawrence: University Press of Kansas.

————. 2008. *A Guide to the Tallgrass Prairies of Eastern Nebraska and Adjacent States.* Lincoln: Digital Commons, University of Nebraska–Lincoln. http:// digitalcommons.unl.edu/biosciornithology/39.

————. 2012. "Birds of the Tallgrass Prairies." *Prairie Fire* 6(7): 16–19. http:// www.prairiefirenewspaper.com/2012/07/birds-of-the-tallgrass-prairies.

Zimmerman, J. L. 1993. *The Birds of Konza: The Avian Ecology of the Tallgrass Prairie.* Lawrence: University Press of Kansas.

Nebraska's City-Dwelling Peregrines

Johnsgard, P. A. 1990. *Hawks, Eagles and Falcons of North America: Biology and Natural History.* Washington DC: Smithsonian Institution Press.

————. 2010. "The Peregrine Falcons of Nebraska." *Prairie Fire* 4(8): 12–14. http:// www.prairiefirenewspaper.com/2010/08/the-peregrine-falcons-of-nebraska.

The Romance of the Yucca and the Yucca Moth

Johnsgard, P. A. 2001. *The Nature of Nebraska: Ecology and Biodiversity.* Lincoln: University of Nebraska Press.

————. 2009. "The Oldest Romance in the West." *Nebraska Life* 13(2): 64–67. http://digitalcommons.unl.edu/johnsgard/42.

A Dazzle of Hummingbirds

Johnsgard, P. A. 1997. *The Hummingbirds of North America*. 2nd. ed. Washington DC: Smithsonian Institution Press.

————. 2009. "A Hummer Summer." *Bird Watchers' Digest* 31(8): 34–39.

————. 2012. "A Dazzle of Hummingbirds." *Prairie Fire* 6(9): 12–13. http://www.prairiefirenewspaper.com/2012/09/a-dazzle-of-hummingbirds.

A Symphony of Swans

Johnsgard, P. A. 1975. *Waterfowl of North America*. Lincoln: University of Nebraska Press. Rev. ed., 2010, Digital Commons, University of Nebraska–Lincoln. http://digitalcommons.unl.edu/biosciwaterfowlna/1.

————. 1978. *Ducks, Geese, and Swans of the World*. Lincoln: University of Nebraska Press. Rev. ed., 2010, Digital Commons, University of Nebraska–Lincoln. http://digitalcommons.unl.edu/biosciducksgeeseswans/.

————. 2013. "The Swans of Nebraska." *Prairie Fire* 7(1): 12–13. http://www.prairiefirenewspaper.com/2013/01/the-swans-of-nebraska.

A Plethora of Pelicans

Johnsgard, P. A. 1993. *Cormorants, Darters and Pelicans of the World*. Washington DC: Smithsonian Institution Press.

————. 2012. *Wetland Birds of the Central Plains: South Dakota, Nebraska and Kansas*. Lincoln: Digital Commons, University of Nebraska–Lincoln. http://digitalcommons.unl.edu/zeabook/8. Also available in hardcopy from Zea E-Books, Lulu Enterprises, http://www.lulu.com/spotlight/unllib.

————. 2013. "A Plethora of Pelicans." *Prairie Fire* 7(3): 9–11. http://www.prairiefirenewspaper.com/2013/03/a-plethora-of-pelicans.

A Gathering of Eagles

Johnsgard P. A. 1990. *Hawks, Eagles and Falcons of North America: Biology and Natural History*. Washington DC: Smithsonian Institution Press.

————. 2011. "The Raptors of Nebraska." *Prairie Fire* 6(11): 14–15. http://www.prairiefirenewspaper.com/2011/11/the-raptors-of-nebraska.

————. 2012. "The Eagles of Nebraska." *Prairie Fire* 6(12): 9, 14, 19, 20. http://www.prairiefirenewspaper.com/2012/12/the-eagles-of-nebraska.

Jorgensen, J. G., S. K. Wilson, J. J. Dinan, S. R. Rehme, S. E. Steckler, and M. J. Panella. 2010. "A Review of Modern Bald Eagle (*Haliaeetus leucocephalus*) Nesting Records and Breeding Status in Nebraska." *Nebraska Bird Review* 78:121–25.

A Parliament of Owls

Johnsgard, P. A. 2002. *North American Owls: Biology and Natural History*. 2nd ed. Washington DC: Smithsonian Institution Press.

———. 2012. "The Owls of Nebraska." *Prairie Fire* 6(2): 15, 20. http://www.prairiefirenewspaper.com/2012/02/the-owls-of-nebraska.

The Feathers of Winter

Johnsgard, P. A. 1998. "A Half-Century of Winter Bird Surveys at Lincoln and Scottsbluff, Nebraska." *Nebraska Bird Review* 66:74–84. http://digital commons.unl.edu/biosciornithology/5/.

———. 2006. "Recent Changes in Winter Bird Numbers in Lincoln, Nebraska." *Nebraska Bird Review* 74:16–22.

———. 2011. "The Feathers of Winter." *Prairie Fire* 5(12): 17–20. http://digital commons.unl.edu/johnsgard/36/.

Johnsgard, P. A., and T. Shane. 2009. *Four Decades of Christmas Bird Counts in the Great Plains: Ornithological Evidence of a Changing Climate*. Lincoln: Digital Commons, University of Nebraska–Lincoln. http://digitalcommons.unl.edu/biosciornithology/46/.

Rich, T. C., C. Beardmore, H. Berlanga, P. Blancher, S. Bradstreet, G. Butcher, D. Demarest, E. Dunn, W. Hunter, E. Inigo-Elias, J. Kennedy, A. Martell, A. Panjabi, D. Pashley, K. Rosenberg, C. Rustay, J. Wendt, and T. Will. 2004. *North American Landbird Conservation Plan*. Ithaca NY: Partners in Flight and Cornell University Laboratory of Ornithology.

A Summing Up

Anonymous. 2013. "Expect More Impacts as Nebraska Drought Rolls into Second Year." *Water Current* (Nebraska Water Center, University of Nebraska–Lincoln), Winter 2013, 3, 13.

Haag, J. 2013. "Thin to Win." *Nebraskaland* 91(3): 34–39.

Johnsgard, P. A. 2007. *The Niobrara: A River Running Through Time*. Lincoln: University of Nebraska Press.

———. 2008. *The Platte: Channels in Time*. 2nd ed. Lincoln: University of Nebraska Press, Lincoln.

National Research Council. 2012. *Climate Change: Evidence, Impacts, and Choices*. Washington DC: National Academy of Science.

Ress, S. 2013. "UNL Scientist Cites Managing Risks in New Keystone XL Pipeline Route." *Water Current* (Nebraska Water Center, University of Nebraska–Lincoln), Winter 2013, 1, 10.

Williams, T. 2010. "Splendor in the Grass." *Audubon* 112 (5): 52–55.